# TSR 2

# Phoenix or Folly ?

Frank Barnett-Jones.

# TSR2 - PHOENIX OR FOLLY?

First published 1994
by GMS Enterprises
67 Pyhill, Bretton
Peterborough
England
PE3 8QQ

Tel & Fax 0733-265123

© Frank Barnett-Jones.

ISBN 1 870384 27 X

The main text of this book
has been set in 10pt 'Times'
on an Apple Macintosh DTP system

All rights reserved.
No part of this publication may be reproduced,
stored in a retrieval system, or transmitted, in any form
or by any means, electronic, mechanical, photocopying,
recording or otherwise
without the prior permission of the publishers.

Printed and bound for
GMS Enterprises by
Woolnough Ltd
Express Works
Irthlingborough
Northants

# Contents

| | | |
|---|---|---|
| Bibliography | | 4 |
| Foreword | | 5 |
| Acknowledgements | | 6 |
| Introduction | | 7 |
| Chapter 1 | Early Days | 9 |
| Chapter 2 | TSR-2 The Plane | 35 |
| Chapter 3 | A Major Step Forward | 63 |
| Chapter 4 | The Olympus Story | 89 |
| Chapter 5 | The Testing Team | 115 |
| Chapter 6 | The Proving Ground | 137 |
| Chapter 7 | Behind Closed Doors | 155 |
| Chapter 8 | People, Planes and Plans | 187 |
| Chapter 9 | The Penultimate Chapter | 223 |
| Chapter 10 | Conclusion | 241 |
| Appendix I | Aircrew Flight Log, TSR-2 XR219 | 247 |
| Appendix II | TSR-2 Production | 248 |
| Appendix III | Contractors | 249 |
| Appendix IV | Important Dates | 251 |
| Apprendix V | Preserving what is left | 253 |

## Bibliography.

| | |
|---|---|
| *BAC* | Charles Gardner |
| *Buccaneer* | Maurice Allward |
| *Crisis in Procurement* | Dr Geoffrey Williams |
| | Frank Gregory |
| | John Simpson |
| *Diaries of a Cabinet Minister 1964-1966* | Richard Crossman |
| *English Electric and their Predecessors* | Stephen Ransom |
| | Robert Fairclough |
| *Fighter Test Pilot* | R.P.Beamont |
| *Jaguar* | Arthur Reed |
| *Modern Fighting Machines F-111* | Bill Gunston |
| *Mountbatten* | Philip Zeigler |
| *Murder of TSR-2* | Stephen Hastings |
| *Oceans Apart* | Cockley |
| *Olympus - The first 40 years* | Alan Baxter |
| *Out of the Wilderness 1963/67* | A.Wedgwood-Benn |
| *Phoenix into Ashes* | R.P.Beamont |
| *Portal of Hungerford* | Derek Richards |
| *Project Cancelled* | Derek Wood |
| *Shorts Aircraft* | C.H.Barnes |
| *The Jet Engine* | Rolls-Royce |
| *The Labour Government 1964-1970* | Harold.Wilson |
| *The Time of my Life* | Denis Healey |
| *Tornado* | Francis.K.Mason |

Extracts from various BAC publications and flight manuals.
Extracts from the following Aviation Magazines.

| | | |
|---|---|---|
| **Air Pictorial** | - | Sept 1981 |
| **Aircraft Engineering** | - | Dec 1963, Nov 1964 |
| **Aviation Monthly** | - | Sept 1973 |
| **Flypast** | - | May 1960, Nov 1989 |

# Foreword

Without doubt the complexity of the advanced Operational Requirement resulting in the design of the TSR-2 meant that it was to the forefront of technology when it embarked on its flight test programme in 1964. Indeed, Tornado pilots who have recently studied the TSR-2 Flight Manual agree that it reads like that of a longer range IDS Tornado.

No military aircraft programme launched since the end of the 2nd World War has provoked greater controversy, and the alleged politico-economic grounds for the cancellation of the programme have been the subject of debate and conjecture over the years.

In addition to being the most complex and advanced military aircraft ever built in the U.K. up to that time, there were the additional problems of the politic amalgamation of the aircraft companies selected to design, develop and produce the TSR-2, (Vickers-Armstrong, English Electric, and Bristol Aeroplane), new working standards and, not least, fierce opposition to the project from certain quarters.

Frank Barnett-Jones has applied himself with dedication and perseverance in covering the TSR-2 story in depth with detailed discussion of the concept, systems design, and proposed flight test programme. He also makes interesting revelations on the political scene, spiced with considered comment on the biased and misinformed statements from so called "expert" aviation journalists.

Aviation enthusiasts and anybody remotely connected with the TSR-2 story will find this book revealing and enthralling, such that in my humble opinion it should be highly recommended reading for RAF Staff College students and politicians of all parties - in fact, anybody involved in the drafting of military aircraft operational requirements and procurement of the resultant hardware.

*Jimmy Dell*

Wing Commander James.L.Dell O.B.E. RAF (retired)

# Acknowledgements.

In writing this book, I required the help and co-operation of many people, especially as it contains such a vast amount of detailed and technical information. Perhaps it was significant that those who originally worked on and flew the TSR-2 were able to offer the most help. In particular, a special mention of thanks to Jimmy Dell not only for his superb Foreword and his own contribution as a test pilot, but also for helping me to locate many of those who were also actively involved in the TSR-2 programme. Without this valuable assistance many doors would have remained closed.

Roland Beamont, the Chief Project pilot on TSR-2 and now himself an established aviation author, gave a great deal of assistance in both time and effort to ensure that the facts behind the testing and development phase of the aircraft were correct. His contributions have greatly enhanced this book.

Those that were involved in the project at the time who were keen to offer their help included Peter Moneypenny, Don Knight, Brian McCann and Len Dean. In particular the help and advice from Peter Moneypenny and Brian McCann in sorting out the complex Nav/Attack systems was invaluable.

The contributions from the men who worked on the Olympus engine was most important. Particular thanks go to Alan Baxter, Bill Barrett, Des Lucy and Richard Elder, who racked their brains trying to remember the details from years ago. Also my colleagues at Rolls-Royce, but specifically to Steve Beeson for helping me get it down on paper, along with Andy Siddon and Eric Brindley in photographic.

Many government establishments have offered varying amounts of help, but a special thanks to Peter Godfrey at the Proof and Experimental Establishment, Shoeburyness for his superbly detailed account of the demise of the TSR-2s out on the ranges. Thanks also to Len Woodgate, now retired from the Aerospace Museum, Cosford, who made available to me various documents on the subject held at the museum. There have been so many faces and so many names and to list them here would require a book in itself, but for the record, the following people and companies have made an important contribution to this book. Jack Bryce, Richard Down, Ian Frimston, George Harris, Jim Henderson, J.E.Pateman C.B.E. F Eng, Wally Rouse, Stuart Scott.   British Aerospace, Rolls-Royce plc, GEC Ferranti, General Dynamics, Lockheed Corporation.

I would also like to thank a very good friend, Phil Kingham. Phil not only inspired the title but also designed and painted the cover. The excellent artists impressions and some of the technical illustrations, are also Phil's work and have certainly enhanced this book.

During my researching I was saddened to hear of the sudden death of Don Bowen. Although at the time Don had not made any contribution, Brian McCann was in the process of talking to him in the hope of securing his assistance. Don, as the Chief Project Navigator on the TSR-2, made an important contribution to the programme. His work was mainly involved with the complicated navigation systems, working from Weybridge. When he retired, Don moved to Canada and it was only just before his death that I had managed to contact him.

Just as the book was about to go to the printers I sadly learned that Peter Moneypenny had passed peacfully away on Tuesday 17th May 1994. I had first met Peter some three years ago whilst researching for this book and I felt most fortunate to have secured his assistance. Although he was still recovering from the effects of a stroke he suffered in 1981, Peter showed a great deal of patience and understanding as he guided me though the complexities of the systems he had worked on that were destined for the TSR-2.

The many achievements of Peter's working life are catalogued elsewhere in the book, and I am sure there will be many tributes from those who worked with him. His contribution, not only to the TSR-2 programme, but to aviation as a whole, was extremely important and although the recognition of his work may not have been given the credit he so richly deserved, I am sure there will be many who will remember him for his tenacity and hard work.

Although I only knew him for a short time, I was touched by the warmth of his friendship and impressed with the determination shown to get on with life. I sincerely hope that his contribution to this book will serve as a lasting tribute to his memory. He will be sadly missed.

# TSR2 - PHOENIX OR FOLLY?

# Introduction

The story of the TSR-2 is a somewhat tragic tale, not only because of the tremendous technical achievement which was wasted, but also because of the cost in terms of human endeavour. Such endeavour was brushed aside by a government that traditionally had always claimed to have the working man's interest at heart. A huge workforce of men and women had been charged with designing and building an aircraft whose blueprint had seemingly been plucked from a science fiction novel.

The TSR-2 was conceived in a period when the Cold-War was reaching a crisis point. The Royal Air Force at the time had no aircraft capable of penetrating deep into enemy territory at low level. The current V-Bomber force had been designed for high level penetration but technology was beginning to show that such tactics were becoming vulnerable to the new generation of surface-to-air missiles.

During the late 50's the Ministry of Aviation had laid the foundations for the development of such a bomber. However, in putting forward their proposals, the government also insisted that the aircraft industry should be prepared to rationalise itself, for the contract would be awarded not to a single manufacturer, but to a consortium of companies. By the early 60's the industry had formed into two major manufacturing establishments, the British Aircraft Corporation and the Hawker Siddeley Group. The aero engine manufacturers had also been given a similar inducement whereby the engine contract for the new bomber would be awarded to those companies merging into a single manufacturer. From this emerged Bristol Siddeley Engines Ltd to take on the might of Rolls-Royce, who had combined with Napier Aero Engines.

The technology alone that was to be built into the aircraft would stretch the industry to the limit. The project had been designed and built with such innovative methods that the eventual cost was to be far in excess of any single programme the British government had ordered before. Opposition to the project had been extremely fierce and forthright from the very beginning, particularly from the opposing political party.

In 1964 the newly elected Labour government, already facing a monetary crisis, embarked on a process that would endeavour to eliminate, not only all Conservative instigated military aviation projects, but also the Anglo-French Concorde. In an attempt to resolve their problems, the Labour government approached the International Monetary Fund, (IMF), hoping to secure a loan in order to avert having to devalue the pound. Sensing its opportunity America siezed on this as a way of pressuring Britain into abandoning its advanced aviation programmes, which included Concorde and TSR-2. This certainly seemed to be the case when the Americans appeared to manipulate a British delegation to the States into accepting a deal involving the new General Dynamics TFX or F-111 as a TSR-2 replacement. In return America would not only 'rubber stamp' the IMF loan, but would also supply Britain with the F-111 with what appeared to be a shallow promise of favourable terms in offset trade.

When the British industry had achieved all that had been asked of it under the TSR-2 programme, the government just literally went ahead and destroyed the programme, leaving the industry on the verge of collapse. The truth behind the decision to cancel the project may never be fully known. Certainly at the time there appeared to have been no in-depth study of the programme itself or any inquiry into whether the country required such a weapon in order to maintain an effective defence system. What followed was seen to be an irrational and irresponsible action which more or less obliterated the project from the face of the earth and, furthermore, the government issued instructions for the complete destruction of anything even referring to the TSR-2.

The real cost of this destruction will never be known, but suffice it to say that this devastating decision has cost the taxpayer literally millions of pounds. Even after the project was cancelled the same government was still in effect admitting there was a need for a weapon of this calibre when it placed an order with the Americans for their new aircraft. Although similar in its concept, the foreign bomber would have failed to meet the requirements of the services in more than one area and this casts serious doubts on the credibility of the information given to the government regarding the technical specification of the foreign aircraft. Ironically, three years after ordering the American F-111, the Ministry of Defence was urged to cancel the aircraft. Not only had the price of the F-111 exceeded that of the TSR-2, but the government maintained the requirement for such a weapon was no longer relevent in the current world atmosphere.

The cancellation clauses for the American aircraft meant that the government had to pay in excess of £40 million pounds in addition to the huge cancellation bill for the British plane. This additional waste of money illustrated how irresponsible the Socialist government had been in managing the whole affair. The Royal Air Force was left with outdated equipment for nearly ten years until the Panavia Tornado joined their inventory.

Frank Barnett-Jones
Nottingham
August 1994

## Author's Note.

The views and opinions expressed in this book are those of the author and do not reflect those of individuals or companies who have contributed to this publication.

# CHAPTER ONE
## Early Days

It was nearly three-thirty on a Sunday afternoon, in September 1964, when a new aeroplane took off from A&AEE, Boscombe Down on its maiden flight. This was the culmination of over seven years hard work, often in the face of tremendous adversity. Although the flight was to last only fifteen minutes, the report submitted by the chief project test pilot stated that it was very successful and had shown few problems, none of which were insurmountable. Over the next seven months three pilots were to make a total of 23 further flights in TSR-2 XR219, before the whole project was cancelled by the Labour government in 1965.

This is a story of success and failure. The success was TSR-2, an aircraft born many years ahead of its time and some say, born in the wrong country. The failure was the government's lack of foresight in backing the technological achievement made by the British aircraft industry.

Before looking at the story of TSR-2 in detail, it may be worthwhile reviewing some of the early events that were to affect the whole British Aircraft industry and ultimately TSR-2 itself. This same industry had of course given a tremendous amount during the second World War and now that peace had been formulated, it seemed an ideal time to review and revitalise the whole industry. Toward the latter stages of the war the Germans had demonstrated their knowledge of missiles and rocketry with the V-1 and V-2 and now, as the 3 major powers seemed set to share in the spoils of victory, it appeared only logical that a review should be made of this type of technology. Accordingly, by 1950 Britain had put together plans for a medium range ballistic missile. However, perhaps the most significant development in aviation between the 30s and 50s must have been that of the jet engine itself and its effect on air transportation and aircraft design. The development of the jet engine was in itself an achievement which brought very

Air Commadore Whittle working on the early jet engine. Whittle's early work was never given the recognition the project deserved, resulting in jet engine development being held back. (*Rolls-Royce plc*)

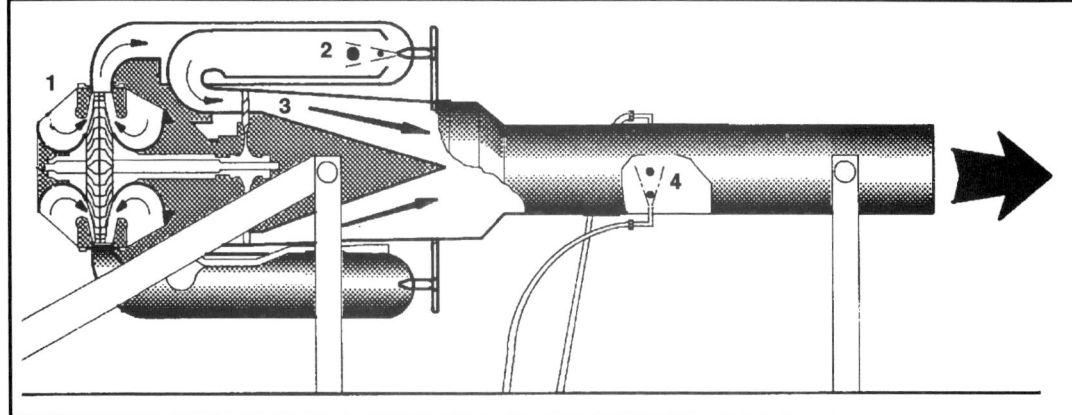

The early W2/700 type engine with a crude form of ATHODYD. Air was drawn in either side of the compressor (1), then forced into the cannular flame tubes (2), before powering the turbine (3). The hot gases exited the jet exhaust into the long tube, where it was mixed with fuel (4) injected into the pipe. *(Phil Kingham)*

little recognition and reward in its early days and most people simply failed to grasp its huge potential which, of course, served only to hold up the development of jet flight.

**Subsonic to Supersonic.**
Shortly after the Second World War, in 1946, the British aviation industry was poised for a breakthrough in the field of supersonics. The technology was there with the development of an aircraft from the Miles Company, the Miles M52. There was a proposal for the aircraft to be fitted with one of the new Frank Whittle W2/700 Power Jet engines, which, in conjunction with a crude form of afterburner, would enable the M52 to fly supersonic.

Thrust Augmentation or ATHODYD (Aero Thermo Dynamic Duct), now known as 'Reheat', was designed as an extension to the jet engine. It was believed that not even the Germans had succeeded in taking their development this far, but ATHODYD was to bring new and unforeseen problems to the development of the jet engine.

It seems most unfortunate that the Whittle engine was never to be used in the Miles M52. The only flight testing being done was by using models of the aircraft launched from an adapted Mosquito. Although overall the results appeared good, the sceptics were becoming worried. There were certain departments that voiced

The Vickers submission for the Experimental Research study. Very similar to the Miles M-52. *(Royal Aerospace Establishment)*

# EARLY DAYS

A model of the M-52 just after launch from the Mosquito. *(Royal Aerospace Establishment)*

their concern over this very dangerous and unknown area of flight and many felt that a more in-depth study should be made before anyone committed any hardware to the sky and possibly risked the life of the pilot on board. The Royal Aircraft Establishment (RAE) at Farnborough, which had been involved right from the start, let it be known that it considered the M52, with its small 5ft diameter fuselage and straight, razor thin wings, unsuitable and definitely not the aircraft to pioneer such a huge undertaking. In February 1946, Sir Ben Lockspeiser, Director General of the Scientific Research, MAP, wrote to the Miles Company saying that it had been decided to cease development of the M52. The whole project had been shrouded in mystery and when it was cancelled the Official Secrets Act forbade any release of information to the public. The Ministry of Supply (MoS) instructed all data and information that had been gained so far to be passed to the Americans, who at the time were deeply involved with the Bell X1. This was a most unexpected, but very welcome gift to the Americans and helped them in their quest to build the first aircraft officially to fly supersonic.

The Bell X1 was a very similar design to the Miles M52. The X1 had straight wings but with a conventional cockpit (as opposed to the M52 with its cramped cockpit in a bullet shaped fairing with the pilot flying the aircraft in the prone position) but by far the biggest difference between the M52 and the X1 was the powerplant. The X1 was powered by a Reaction Motors E6000-C4 rocket motor, whereas the M52 was to be powered by a conventional jet engine.

**Political Adversity.**
In 1946 Barnes Wallis started work on variable geometry technology whilst working at Vickers. His first work in this particular field was the Wild Goose and whilst working on this project he introduced the Swallow, again a variable geometry design. The Mutual Weapons Development Programme (MWDP), a joint venture which included the United States, became very interested in the Swallow, for in 1951 Bell Aircraft had flown their own variable geometry design, the subsonic Bell X-5. This particular aircraft was a direct derivative of the German Messerschmitt P1101 appropriated by the Americans just after the war. The Swallow was in fact far more advanced than the Bell X-5 and after the Sandys'

# TSR-2 PHOENIX OR FOLLY?

White Paper of 1957 cancelled the project, it was no surprise when the Americans showed renewed interest. After an agreement between the MWD, the British government and Vickers, a six month programme was agreed on to develop a manned variable geometry aircraft. America were given all of the Barnes Wallis data and although the project was plagued with problems, it gave America what they wanted and with this information they eventually built their variable geometry fighter/bomber, the F-111.

The war years had taken their toll on the development of new types of aircraft, as the overriding priority had been simply to get new aircraft out of the factories and into the war. Understandably, very little time and effort had been available to develop new or controversial projects. Therefore it had been the aircraft currently in service, the backbone of Bomber Command, which had to be modified or further developed in order to carry out any specialised roles demanded by the RAF. By 1944 the Avro Lancaster had been so redesigned that it became in fact a new bomber known as the Avro Lincoln. The Lincoln went on to be Britain's front line bomber right into the early 50's. Although it was an admirable aircraft, the advent of the nuclear bomb amply demonstrated the fact that by 1947 Britain was trailing behind America in new aircraft development, with the USA re-arming with new, faster bombers fitted with jet engines that could fly at far greater heights.

The question is, did the British government fully realise just how far behind Britain was falling - in bomber development alone - especially when compared to the United States and the Soviet Union? In Britain, at the time, there was only a single parallel development which was the B3/45 English Electric Canberra. The government, while feeling that the threat of war was very unreal so soon after World War 2, appears to have been caught off guard with the outbreak of the Korean War in 1950. This led to an immediate investment for the development of new and better aircraft.

The whole programme was, however, hindered by poor facilities and material shortages. RAF Bomber Command was to benefit the most from this reversal in Government policy, when a programme was embarked upon to re-equip Bomber Command with aircraft powered by jet engines that would be able to compete in performance with the new American bombers. Contracts were issued to various manufacturers to build these new aircraft, which were to be known as the V-Force. The Valiant, Victor and Vulcan were to serve with the RAF into the late 80's and early 90's. The Valiant was only to serve for a short period due to severe problems with its airframe.

The Avro Lincoln was the backbone of Bomber Command right into the early 1950's before being replaced by the V-Force.

(Rolls-Royce plc)

# EARLY DAYS

The MK 5 Supermarine Swift. A controversial airplane pressed into service before it was ready. The resulting inquiries imposed tremendous restrictions on future military projects. *(Rolls-Royce plc)*

**Politics and Planes.**

Having by far the most profound effect on aircraft procurement in the period just after the war must have been the affair surrounding the Supermarine Swift. Not only were the MoS and the industry itself in a state of confusion, but also it seemed, were the RAF. At the time the Technical Directorate of the MoS appeared to have suffered very little from any restriction of its ability to order military aircraft. As a consequence it was apparent that little thought or strategy had been applied when orders were placed with the manufacturer. In the late forties it was deemed necessary for the RAF to have a fighter with a very high rate of climb and a performance in excess of 600 mph once it reached its ceiling. The competition was narrowed down to the Hawker Hunter and the Supermarine Swift. In terms of development Supermarine were slightly ahead of the Hawker camp. They had already gained considerable experience with the naval Attacker and as a result Supermarine won the order.

Right from the very beginning it was obvious that too much was being asked of the aeroplane in too short a time. This resulted in the project becoming badly affected by a string of technical problems. 150 aircraft were ordered and pressed into service far too soon. Modifications were still being applied as aircraft were being delivered. In 1951 an assessment dismissed the possibility of cancelling the Swift, apparently because of the huge sums of money which had already been spent on the project. It was to be four years before the axe finally fell. In 1955 the media were informed of the government's intentions at a hurriedly convened Press conference. Harold Macmillan, then Minister of Defence, and Selwyn Lloyd, Minister of Supply, attempted to explain the government's actions to the hostile media. Their brief was badly prepared and ill informed so when the media put forward their questions, the ministers could say very little because of their lack of basic knowledge. The Press were outraged at the ineffectivness of the process by

It seems that Hawkers attempts to break into supersonics appear to have been thwarted in every case, possibly as a consequence of the Swift affair. Further development of the Hawker Hunter was cancelled by the RAF. *(Rolls-Royce plc)*

which the government procured military aircraft. Not only did the affair alienate the media and the public, but it embarrassed the government. The whole subject became the study of a White Paper, but the damage had been done. Many in government blamed the industry, accusing it of being grossly inefficient and producing shoddy goods. The effect of this unfortunate epic served only to tarnish the reputations of future projects.

Up until this time there appeared to be no satisfactory method by which the various projects undertaken by the aircraft industry at the behest of the Ministry could be properly co-ordinated to achieve exactly what was required by the Armed Forces. Indeed it was felt that at this time without more 'control', such projects ran the considerable risk of producing an end result altogether different from what the Ministry originally intended. The job of the Operational Requirement Staff (Air), was basically to determine the future weapons systems of the service and set them down in a document known as an 'Operational Requirement'. The O.R. was then studied and refined usually resulting in a technical specification which would, in turn be shown to the industry. From this a feasibility study would then be made of the project which would hopefully lead to project development and the building of a prototype. This was, of course, for the successful projects, but it was not always the O.R office who raised particular projects. Occasionally a manufacturer raised a program of its own and submitted it to the O.R. office for consideration. Today the functions of this office have been centralised, especially as the procurement of military hardware is made all the more difficult by the number of multi-national collaborations.

The apparently haphazard way of project control was finally brought to the attention of the Ministry of Defence (MoD) by a White Paper prepared during 1955/56 on the supply of such aircraft and the research being done in this field. At this time it was solely the Defence Research Policy Committee (DRPC), that could decide the priority of the projects forwarded to them by the Operational Requirement office. The apparent lack of technical expertise in the DRPC very often led to projects of a similar nature or with similar technical objectives being evaluated in parallel, thereby wasting the industry's resources and efforts. During 1956/57 stringent changes were made to ensure the procurement of military aircraft had a more sound basis. This was achieved by the inclusion on the DPRC of senior, technical RAF officers who would serve for longer periods, than the normal 2 years. This was intended to provide more continuity in project development and would hopefully eliminate confrontation caused by differing opinions. Another

The English Electric B-2 Canberra. Conceived in the late 40's, the Canberra in various marks, has not only been one of Britain's most successful exports, but also one of the RAF's longest serving aircraft. *(Rolls-Royce plc)*

# EARLY DAYS

recommendation was to ensure that any new military aircraft should be considered as a complete weapons system and not just an armed flying machine flown by a pilot. Consideration was also to be given not solely to its potential, but to the very mechanism by which that potential could be used to its best effect. Thus, the airframe and all equipment required was to be considered as a whole. A further recommendation was the adoption of the development batch system. This meant that a small number of prototypes would be built, for example a total of twelve. It would ensure that the development stages would not be held up due to a shortage of aircraft. The normal process of building one or two prototypes often led to bottlenecks in development because of the shortage of aircraft. Therefore having a greater number of aircraft would ensure a quicker development path.

**Initial Steps.**
There was however, one particular company that appeared to be going from strength to strength, seemingly quite unaffected by the spate of Government cuts prevalent at the time. This was the English Electric Company. It was during the first part of 1944 that the idea of a jet engined fighter-bomber had been first suggested. English Electric had become involved when W.Petter joined them at their premises in Preston, Lancashire, that same year. By 1949 the new prototype aircraft had been built. It was a twin engined, long-range, high-altitude, highly manouverable bomber, carrying a crew of two, and on Friday the 13th of May, (a date the Test Pilot, Roland Beamont, was heard to say, 'was as good as any other'), the Canberra made a very successful first flight from the English Electric aerodrome at Warton. This was to be a most successful aircraft that sold to many air forces throughout the world, including the USA, where it was built under licence by the Glenn Martin Company and known as the Martin B57. No doubt English Electric considered themselves to be fortunate in having the successful Canberra and at a time when they were also working to establish themselves in the supersonic market. Since the Miles M52 had been cancelled, English Electric had begun studies in supersonics during 1947/48, leading to the F23/49. On the day before the Canberra made its first flight, English Electric received a contract from the MoS to proceed with its design work on an aircraft then known only as the P1.

This aircraft first flew in August 1954 in the hands of Roland Beamont and quickly established itself as an ideal basis for the RAF's proposed high altitude interceptor. The P1 was capable of high supersonic speed and in fact on only its third flight, exceeded Mach 1. This was some time before the controversial Fairey Delta 2. Later the P1 was developed further and became famous as the English

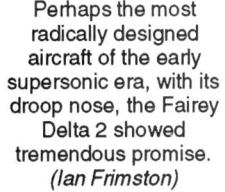
Perhaps the most radically designed aircraft of the early supersonic era, with its droop nose, the Fairey Delta 2 showed tremendous promise.
*(Ian Frimston)*

# TSR-2 PHOENIX OR FOLLY?

Electric Lightning. Whilst it had a significant amount of success as a Mach 2 fighter, the full potential of this aircraft was never fully realised due to what could only have been the short-sightedness of the British Government. The Lightning, in its various Marks, continued in service with the RAF up until 1988, when it was finally withdrawn. Outdated as it was and lacking many of the latest technological refinements, in performance and manouverability, the Lightning was still more than a match for its modern counterparts.

In the early 50's, Fairey Aviation were making a study of supersonics, and they eventually produced the Fairey Delta 2. This was an advanced design aircraft with a tailless delta wing configuration powered by the Rolls-Royce RA501 Avon. The Fairey Delta 2 first flew in October 1954, with Peter Twiss at the controls. Whilst the government failed to support a world record attempt, the world of aviation was amazed in March 1956, when the FD.2 achieved a speed of 1,132 mph shattering the world speed record by 300 mph! This feat caused the United States airframe manufacturers quickly to review their own airframe development. The French, baffled as to why the British Government had decided not to support the FD.2 project at the outset, completely altered their military programme, and, after having closely studied the FD.2 began producing the now famous Dassault Mirage.

**Defence in Jeopardy.**
During the early part of 1957 Duncan Sandys, the then Defence Secretary, presented his White Paper on Defence. The paper seemed to be based on his research done in 1944 when Winston Churchill appointed him to carry out a study on the effectivness of the German missiles and rockets with particular emphasis on the sort of problems these types of weapons could pose. His statement, with its many controversial views on Britain's defence, was to set the development of British military aviation back many years. The whole future of the Royal Air Force and the role it would play in Sandys' thinking was to have a huge demoralising effect on the service. The recommendations were for the Air Force to be equipped with missiles and rockets, with pilots needed purely for transport aircraft only. Manned flight in aircraft such as fighters and bombers was to become a thing of the past. Britain's defence would no longer be reliant upon fighters, but missiles strategically placed throughout the country to ensure incoming enemy bombers would be

Built by the Bristol Aeroplane Company, the Bristol 188 was to explore the effects of kinetic energy to help with the AVRO 730 project. It proved to be a failure mainly because of the unreliability of the Gyron engines and its excessive fuel consumption.

*(Rolls Royce plc)*

# EARLY DAYS

destroyed before they could offload their nuclear bombs. A retaliatory attack would be carried out by rockets capable of flying up into the ionosphere and at thousands of miles an hour. No doubt cost was one of the major influencing factors, with such deterrents being considerably less expensive to develop and manufacture than a jet bomber with the complexities of its power plants and avionics systems.

It was, however, a matter Julian Amery was to contradict. Perhaps the missiles were sufficient to protect the country, but it only needed one bomber to get through and release its bombs to create havoc, causing widespread death and reducing such missile sites to rubble. Then there was the question of retaliation. A missile or rocket could not carry out a reconnaissance, or, photograph potential installations, and a missile site was far less flexible than an aircraft, as any thought of trying to move it would have proven far too uneconomical. He therefore felt the Air Force were justified in demanding an aircraft to carry out these roles.

Even the Operational Requirement office appeared to favour the missile technology expounded in the Sandys' paper, and virtually rubber stamped the cancellation of the proposed Avro 730. This was perhaps one of the first victims of the Sandys' thinking with the project being cancelled in favour of the Blue Streak rocket, a development that itself was to be cancelled in favour of the American Polaris system for the fleet of new nuclear powered submarines for the Royal Navy. The Avro 730 was to have been the worlds first Mach 2+ bomber and at the time contained very advanced thinking that would have eventually helped with the future development of supersonic transport. An offshoot of the 730 was the Bristol 188, designed to assist with its development. The 188 first flew in April 1962, but proved to be a very costly project, due mainly to it being constructed of stainless steel, and the unreliability of its Gyron engines. After a cost of some £20 million, the project was cancelled, having provided no real benefits, and particularly no information regarding the behaviour of 'kinetic energy' at high Mach numbers.

**The Age of the V-Force.**
Britain's V-Force, the Vickers Valiant, Avro Vulcan and Handley-Page Victor began demonstrating their effectivness during a number of multi-nation exercises. The Valiant, Britain's first V-Bomber, first flew on the 18 of May 1951 and by 1955 was entering RAF service. The huge, Delta winged, Vulcan first took to the air in August 1952. The Avro company had gone to great lengths to ensure the integrity of the design by building a number of smaller aircraft using the Vulcan's basic shape, in order to perfect its design before finally building the Vulcan bomber itself. Handley-Page, were similarly concerned with their radically designed Victor and also built a small prototype aircraft again to prove the Victors final design before it first flew on Christmas Eve 1952, from Boscombe Down. The trend of building small development aircraft had been actively encouraged by the

Although Britains first V-Bomber the Valiant was to serve for only a short period due to metal fatigue in the main wing spar.

*(Rolls-Royce plc)*

# TSR-2 PHOENIX OR FOLLY?

government in order to keep the various design offices busy, in an attempt to retain the technology we had. The Vulcan was to carry the new Blue Steel Mk.1 which had the Yellow Sun Thermonuclear warhead with inertial guidance system, though it was limited to a range of 100 miles. Although the Mk.1 entered service with a number of squadrons, a Mk.2 version of Blue Steel had already been proposed with a far greater range of 1,000 miles. However, following a series of lengthy delivery delays, it was cancelled in favour of the American Skybolt.

Skybolt itself was dogged with a number of problems, and was eventually cancelled by the Americans themselves. There had been concern as to the weight of the missile and whether the B-52 wing could sustain this weight for long periods. A reassessment concluded the B-52 would require major and expensive modifications in order to carry it.

In 1962 a meeting was hurriedly convened in Nassau between the Prime Minister, Harold Macmillan and President John.F.Kennedy, Britain was told that the project was to be cancelled. Britain insisted on leaving the meeting adamant that a deal be concluded for a new missile system. Kennedy offered to assist Britain by allowing the development of Skybolt to continue with Britain paying for half of the costs. Macmillan, apparently on the advice of the Chief Scientific Advisor, Solly Zuckerman, turned the offer down because of the huge expense of the project. Lord Mountbatten had already suggested to Harold Watkinson, Minister of Defence, that he should spend the £100 million saved on Skybolt, on two American built Polaris submarines. Instead Britain agreed to purchase the new Polaris missile only and spend further huge sums of money building a fleet of nuclear powered submarines capable of using the system.

Because of an urgent need for an air-to-air refueller the Valiant was converted to take on the role before the metal fatigue problem was discovered. Here a Valiant is topping up a Vulcan B1-A, Britain's second V bomber.
*(Rolls-Royce plc)*

Britain's third V-bomber, the Victor. This was another of the V-bomber types whose role was to change in the latter part of its service life. This Victor is carrying out In-Flight refuelling trials with one of the development Tornado's.
*(Rolls-Royce plc)*

## A Change of Role.

In 1963, Britain's V-bomber Force was assigned to NATO and this called for a new role. The aircraft had been specifically designed for high level bombing but NATO wanted a change in tactics and required them to become low level bombers, mainly to avoid detection by the Russian SAM missile systems.

A severe blow was dealt to the V-Bomber Force, when in August 1964, the crew of a Valiant, flying at around 30,000ft, heard a loud bang in the aircraft. As they were flying in cloud they believed they had hit another aircraft, and decided to return to base. The aircraft appeared to be flying normally and, having burnt off sufficient fuel to make a safe landing, they started the approach. When lowering the flaps, the aircraft started to roll, and the flaps were immediately raised. They continued the approach as normal minus the flaps, but after a perfect landing, the starboard wing appeared to be sagging. The aircraft was moved to the hangar and stripped down for inspection. The Valiant had two large, forged, beams that acted as wing spars, which were also connected to branches that formed part of the immense internal structure running the whole length of the aircraft. It was found that the rear beam of the structure had fractured to the extent that a hand could be placed in the crack! All Valiants were checked for this weakness and fatigue of varying severity was found in this particular beam. The beams were of a new, light-weight metal, and it seems that this metal led to the downfall of the bomber, for it appears very little, if any, testing had been done to establish the 'life' of this new material, (unlike today's materials that are tested to destruction). Vickers had been originally instructed to build a 'stop-gap' bomber, and although this they had achieved, in December 1964, the Valiant flew its last sortie before the entire fleet was scrapped.

## The Seeds are Sown.

In 1955 the Canberra was still serving well, but despite this the Air Staff in the Operational Requirement office felt the RAF would require an aircraft, not necessarily as a direct replacement for the Canberra, but one that would carry out roles that so far the Canberra had not been performing. Originally the O.R office had felt the English Electric aircraft had been built only as an interim bomber, and so it was understandable that sooner or later their thinking would turn in this direction. With the latest proposals on the future needs for the RAF, and in the light of the Sandys' report, a situation where the Air Force would be entirely dependent upon missiles and rockets was causing concern. On top of this, other major concerns within the industry were the huge cuts in spending on manned aircraft research and the various contracts that were being cancelled in order that monies could be re-directed into developing and building the new missiles and rockets.

## Operational Requirements OR.339.

By March 1957 the O.R. office duly formulated a requirement for an aircraft that would replace the Canberra and the specification was issued under the General Operational Requirement number 339. The first draft of this document was issued by the MoS in October 1957, to a number of selected aviation manufacturers, (see Appendix 4), who were invited to assess the GOR and submit their ideas for consideration.

The issuing of the GOR and responses to it would give the MoS not only an indication as to the state of the industry, but also as to how advanced their thinking was in terms of design and equipment. Although not generally known at the time, it was strongly hinted that the government would prefer this to be a joint project between two or more companies, a move that would perhaps inevitably bring about the demise of certain companies, eventually reducing the amount of aircraft manufacturers to a select few. That 'strong hint' indeed was to become a condition of awarding the contract.

To many of those manufacturers this idea did make sense, especially as there was so little business going around. It was estimated that just after the war Britain had more aircraft manufacturers than the whole of the United States, a situation that was becoming increasingly unacceptable not only to the aviation industry itself but also to the Government, who after all did issue the contracts for all military equipment. The idea was not new, for in 1954 the then Under Secretary, Denis Havilland, had written a paper on this subject. Furthermore it had also been discussed on a number of other occasions especially when it was realised that the industry was only being kept going because of the aircraft attrition rate from the Korean War. It was also believed that as the newer aircraft became more sophisticated, the demand would drop, not only because of cost, but also because the aircraft could amply achieve more with its sophisticated equipment, becoming in effect a Multi Role Aircraft.

Britain's defences at the time were being stretched to their limit and the RAF had no Tactical bomber capable of penetrating enemy radar defences at tree top height to carry out a conventional or nuclear attack. The Air Staff were becoming extremely worried and saw GOR 339 as a last effort to maintain a Bomber Command. General Operation Requirement 339 specified that it should :-

- have the ability to penetrate defences such as radar, fighters and surface-to-air missiles, specifically in Europe where these defences are spread out in varying depth.
- be able to deliver a variety of weapons, including nuclear or steel bombs, with a very high degree of accuracy. It should also be able to carry out photographic or radar type reconnaissance in all weathers.
- have a good radius of action and be able to penetrate deep into enemy territory. This would require the aircraft to fly low and fast to avoid enemy radar.
- have a good ferry range at high altitude and be able to use reliable specific bases in time of war or conflict.
- be able to do a high altitude, high-speed, dash with extended range.
- be able to operate from semi-prepared fields and runways and must be self sufficient in the field.

These are broad descriptions of the GOR and the Operational Requirement in its original form was to change no less than four times. In support of OR339, a more detailed requirement was to be issued under the guise of OR343, following the somewhat greater demands that were to be made on the aircraft.

**Naval Adversity.**

In 1955 the Navy had ordered a number of the new Blackburn aircraft, the Buccaneer. Having studied the Operational Requirement, Blackburn could not see why, with a number of modifications, the Buccaneer itself could not fulfil the role specified by OR339. This solution was also favoured by the Chief of Defence Staffs, Lord Mountbatten, who tried to persuade the RAF to opt for a suitably modified Buccaneer. This would certainly have helped with development costs, thereby making it a less expensive aircraft for the navy and ultimately the RAF. At the time Mountbatten had a great ally in the Treasury, as they too were concerned about the amount of resources being spent on defence projects. This was not the last time Lord Mountbatten was to have an influence on the future of the aircraft that

# EARLY DAYS

The Blackburn Buccaneer was designed originally as a naval aircraft. The attempts by Lord Mountbatten to force the RAF to accept this aircraft as a TSR-2 contender made Mountbatten an unpopular man in certain quarters. This S2B shown at RAF Waddington. *(Author)*

would ultimately be selected to fulfill OR339.

The RAF insisted that the Buccaneer had been designed mainly to fulfil a role at sea and the electronics equipment had been designed to pick up warships, (especially the Russian Sverdlov class cruiser) as well as specific land based targets. Also its bombing capability was considered to be quite inadequate in terms of the pin point accuracy, demanded by the Royal Air Force, for both its nuclear and conventional roles. The RAF required a supersonic aircraft with a terrain following capabilty. The Buccaneer was neither. It also had a very limited range, and it required too long a take-off run. The Mk 1 Buccaneer had Gyron Junior engines, but these proved to be somewhat underpowered for the Buccaneer airframe and were later replaced by the more powerful Rolls-Royce Spey. In 1962 Blackburn submitted a further proposal for OR339. This involved still using the Buccaneer, but with a new thinner wing, an updated radar system, reheated Spey engines and a bogie undercarriage, but was turned down on the advice of Sir Solly Zuckerman due to cost. Zuckerman was another leading figure who was against the idea of spending huge amounts of money on a new aeroplane for the RAF, but not necessarily for the same reasons as Mountbatten, and this is studied in more detail later on.

**Taking Strides.**
As the manufacturer of Britain's first jet bomber it seems logical that English Electric should be considering its replacement at a relatively early stage and, in October 1956, they had initial discussions with the MoS on such a proposal and together appeared to have agreed on the basic structure for the type of aircraft that would be required. At the time some consideration had been given to the aircraft being a Strike Fighter with a performance around Mach 1.3 and radius of action of 350 nautical miles. A feasibility study was made into the prospects of developing the P1-B to perform such a role, but was eventually dropped as the Ministry felt a P1-B derivative would be unsuitable for the necessary tasks and that a totally new aircraft would be required. At the time Ray Creasey was Head of Aerodynamics and New Project Design at English Electric, and the major influence on the TSR-2 design. B.O.Heath, the Chief Project Engineer at English Electric, had made some preliminary designs on a Canberra replacement by the end of October 1956. One of his designs was of an aircraft with a straight shoulder wing below which were suspended the podded engines. Being free of engines, the fuselage was very slim with facilities for a crew of two sitting in tandem. The tailplane was mounted high on the fin. In November the Ministry, having had further discussions with the Air Staff, produced a new specification which included new performance figures.

Consideration was now to be given for the new aircraft to take on a number of

# TSR-2 PHOENIX OR FOLLY?

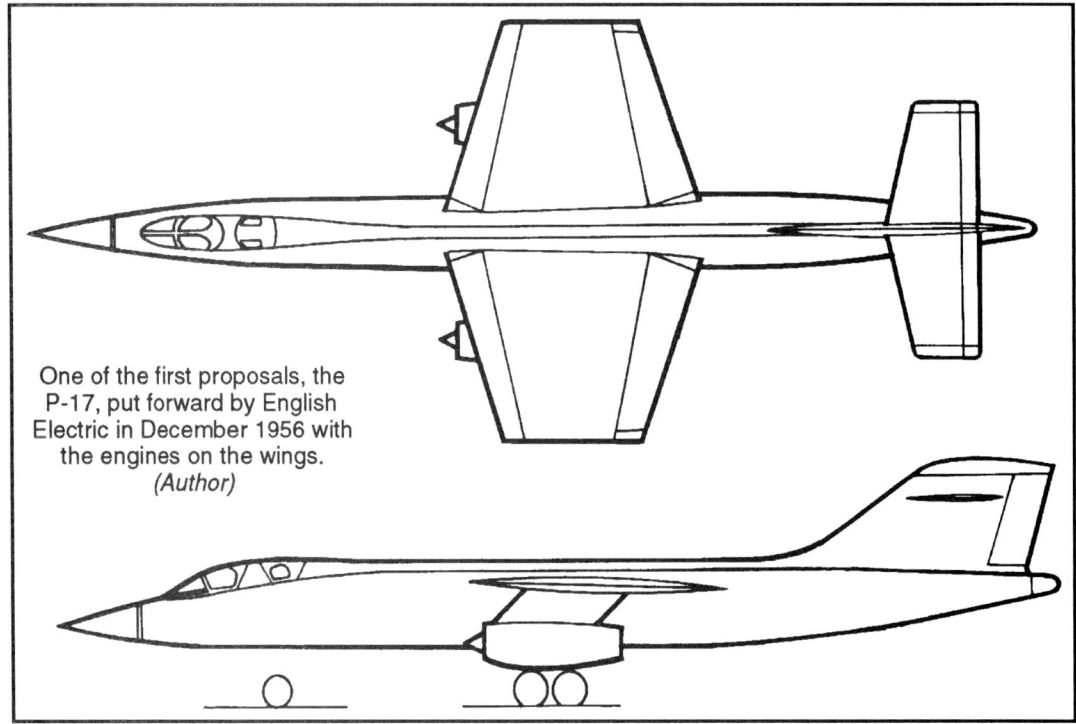

One of the first proposals, the P-17, put forward by English Electric in December 1956 with the engines on the wings.
*(Author)*

roles as well as a new level of performance. It was to have a 1000 mile radius of action with speeds up to Mach 1.5 at altitude, and a radius of 600 nautical miles at Mach 1.3 at sea level. It should also have a high dash capability over the target area with a ferry range of 2000 nautical miles. Apart from the need to carry a variety of weapons, there was also a requirement for the aircraft to take on the role of reconnaissance and therefore carry photographic and electronic reconnaissance devices. The subject of vertical take-off was also mentioned within the requirement in an effort to move away from the reliance on hard runways. By December of 1956 the design teams at Warton were doing low speed wind tunnel tests on an aircraft now designated P-17.

Heath's other design was for a similar straight wing aircraft with highly swept foreplanes. The engines in this aircraft were to be RB133's, as opposed to his other design, for which it was intended to use the RB134. The location of the engines had moved to inside the fuselage, with engine intakes just to the rear of the leading edge: the tailplane had also been moved to mid-fuselage.

Because of the major changes to the initial design, the aircraft was redesignated P-17A and work continued along the guidelines proposed by the Ministry, with the exception of vertical take-off. By mid February 1957, English Electric had been told the in-service date of the new aircraft was to be 1964. The company had already prepared a report, entitled 'Possibilities for a Multi Purpose Canberra Replacement'. The report (based around the P-17), was finely detailed and emphasized that any reliance on a defence system based on rockets and missiles would be putting the country's defence in a near predictable situation with fixed inflexible missile sites. Whilst acknowledging that the costs of such an aircraft in time of economic pressure were of paramount importance, it went on to point out that the company's considerable experience gained from successful projects such

# EARLY DAYS

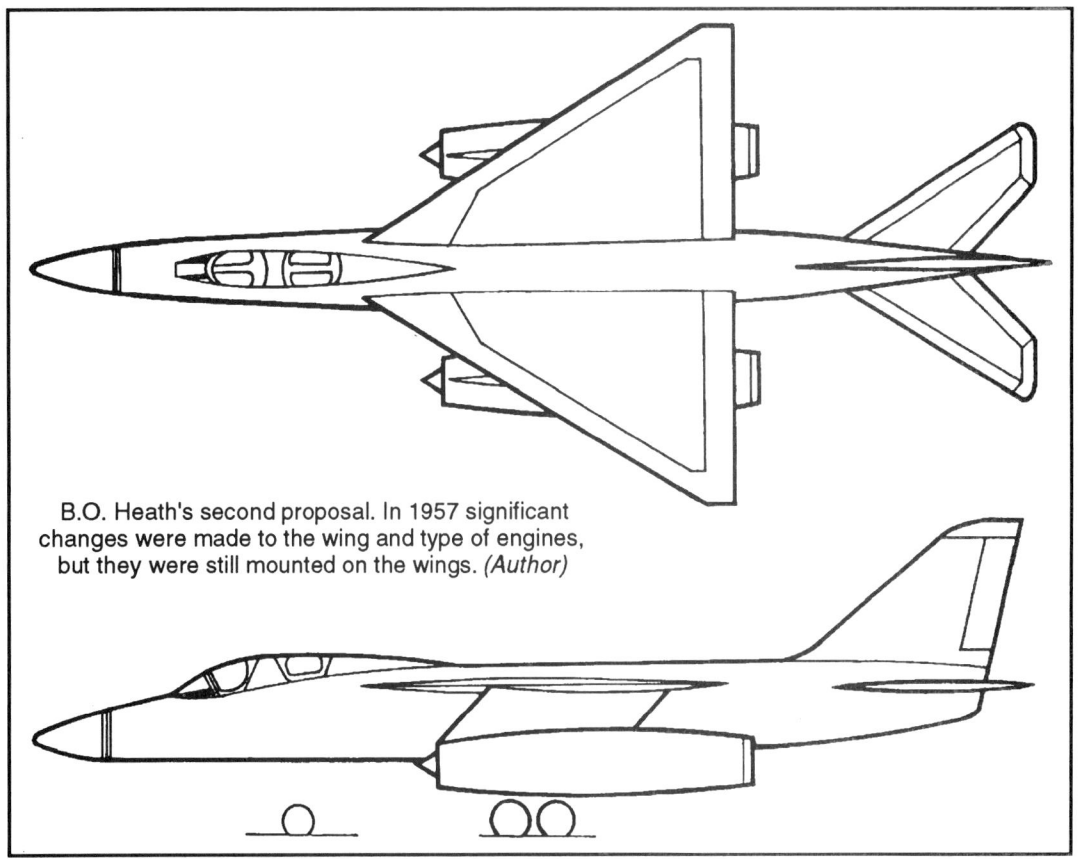

B.O. Heath's second proposal. In 1957 significant changes were made to the wing and type of engines, but they were still mounted on the wings. *(Author)*

as the Canberra and P1, would prove to be a great advantage in the development of the new aircraft. It mentioned the company's success from exporting its products and was confident that an aircraft with the capabilities of the P-17A would also be a contender for similar success.

At the time a Parliamentary Select Committee on Estimates were reviewing the White Paper on Development and Procurement of Military Aircraft and preparing to initiate some of the proposed changes. The paper was presented to Parliament and it was here the first mention was made of the possibility of the RAF accepting early versions of the P-17A, even though they would not be fulfilling a formal Operational Requirement. It also recommended that the government would be wise to restrict the amount of contracts it issued in order to encourage the industry to make better use of its resources by perhaps amalgamating some of the companies. A few days later Duncan Sandys became the new Defence Minister and was soon to present his controversial White Paper on defence that was to have such a profound effect, not only on the RAF, but also on the aviation industry, an effect that was to put the development of British aviation back 10 years.

With the Air Staff having completed their studies in March, it was only left to the MoS to issue the first draft of the General Operational Requirement and this was done later on in October. In the meantime between May and December, English Electric continued the research and development programme for the P-17A design. During May they were testing a slow speed one-eighth scale model in the wind tunnel. Over the next few months various changes were made especially

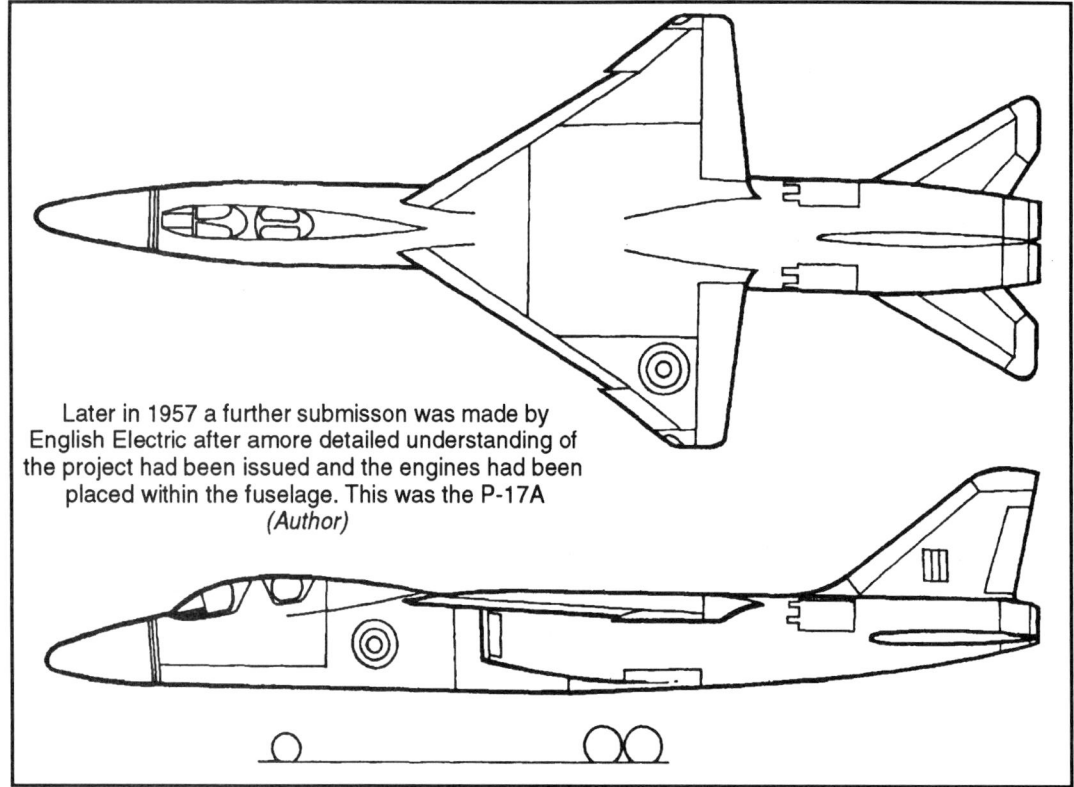

Later in 1957 a further submisson was made by English Electric after amore detailed understanding of the project had been issued and the engines had been placed within the fuselage. This was the P-17A
*(Author)*

to the wing and the need for larger engines was considered. In July and August two further OR's were issued, one for navigation, bombing, reconnaissance and flight control systems, OR3044, and OR3596 for Linescan navigation and reconnaissance systems. The company's involvement in the very early stages of the Operational Requirement gave it a distinct advantage over its rivals; it was however deficient in one important area, and that was the need for Vertical take-off performance. English Electric were unable to develop the P-17A with this capability in the time specified, as it was still scheduled to be in-service by 1964. So, in accordance with the government's thinking on amalgamation within the industry, English Electric began a collaboration with Short Bros of Belfast.

**Close Cooperation.**
This was not the first time the two companies had closely cooperated together. In 1954 Shorts had become involved in the production of English Electric's Canberra, when they obtained a manufacturing contract for 150 of the type. Shorts also worked on Canberra PR3 conversions and converted a number of Canberras to drones. The second time the two companies worked together was during the early development of the P1. Farnborough had concerns regarding the configuration of the main wings in relation to the tailplane on the aircraft. As a result of this, in 1950 the MoS awarded Short Bros a contract to build a small experimental aircraft, using the basic shape and wing layout of the P1, known as the SB-5, which would be used to test out Farnborough's theories. It was subsequently found in test flights with the SB-5 that English Electric had got their sums right.

Shorts were indeed about to start work on an aircraft, known as SC1, capable of

# EARLY DAYS

Shorts were awarded a contract to build a small, slow speed experimental aircraft to prove the design theory of the E/E P1. The SB-5 was built with two tail sections, one (above) with the tailplane on top of the fin and a second (below) with it in the fuselage. The design only confirmed that E/E had done it right. *(Short Bros)*

taking off vertically and because of Shorts experience in this field, English Electric began cooperation on a lifting platform designated PD-17. (PD17 was a Short Bros designation, which meant Preliminary Design, English Electric allocated it P-17D.) This was to be a large delta flying wing with 60 RB108 lift jets, of which 44 were fixed, 16 tilting, and an additional 10 jets for forward movement. The delta wing formed the base for the P-17A to be mounted on and along the leading edges of the wings were the engine intakes. The single cockpit was situated on the centre line at the rear of the aircraft. Just forward of the cockpit was a huge hook to secure the P-17A. The PD17 weighed some 75,000lbs and with P-17A aloft, 150,000lbs. Theoretically this would lift P-17A into the air before the aircraft launched. This joint collaborative proposal was submitted for consideration in a number of substantial volumes, and at the time was the only submission that covered, fully, the whole scope of the Operational Requirement. After an initial assessment by the MoS and MoD the P-17A/PD17 was judged to be the only serious proposal.

English Electric began talks with Short Bros as this was the only company involved with vertical takeoff. The Short SC-1 was designed as an experimental vertical take-off platform. *(Short Bros)*

# TSR-2 PHOENIX OR FOLLY?

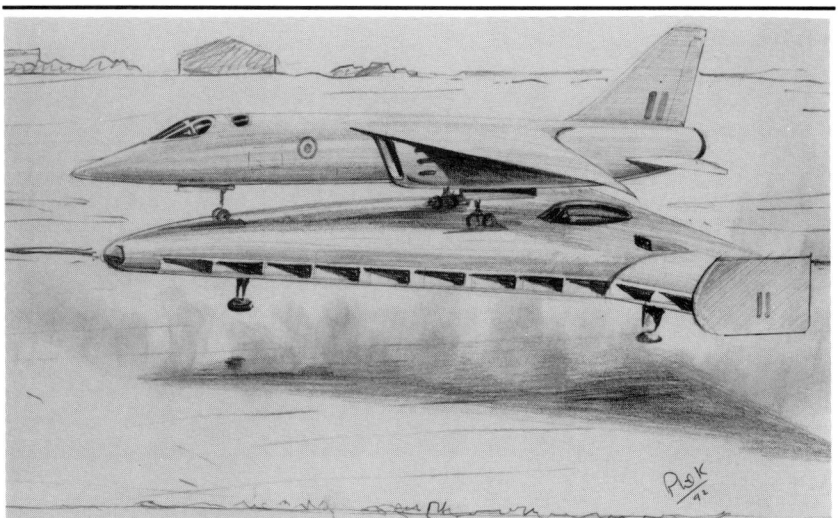

E/E - Short Bros attempt to conform with the VTOSL aspect of OR339. The engine combination in Shorts PD-17 was quite daunting. Never had so many engines in such configurations been proposed in one aircraft.

*(Phil Kingham)*

**A New Association.**
By March 1957 a report had been made on the selection of firms who would be invited to tender for OR339, which was a more detailed study of GOR339. After the detailed submission by English Electric and Shorts, it was strongly felt that these two companies would join together to produce a combination suitable for the OR requirement. As part of the same group, Hawkers and Avro had already agreed in principal, that the two companies would work together. In 1957, Sir Sidney Camm had experimented with the basic design of the Hawker P1121 by putting two engines side by side in the fuselage and designating it the P1125. So when OR339 was issued, Hawkers were urged to submit this design, with certain modifications, as a competitor. The most important modifications were the addition of a rear seat for a navigator, and the introduction of two Rolls-Royce RB141's instead of the original RB133's. On paper this appeared to be an excellent design and with a number of further changes it could well have been a serious competitor for the OR339 contract. However, because of submissions made by companies associated with the Hawker Group, namely Avro and Glosters, for the same contract, the Air Staff had apparently become confused and in fact appeared to defer all Glosters submissions for clarification. The following year the Hawker group did make a single submission to clarify its position and, although based mainly on the original P1125 design, it did include some of the more advanced points of the Avro and Gloster submissions.

Vickers Supermarine of Weybridge on the other hand, had been unable to submit a detailed design in respect of GOR 339 and at the time were deeply concerned as to the consequences of being omitted from this most important military project. After having talks with the powers in Whitehall it was strongly hinted that Vickers should have exploratory talks with English Electric on the subject. In April 1958 talks began between the two companies to establish a basic idea of what equipment such an aircraft should have. It was after these talks that in June of that year the MoS made its final assessment and a short list was drawn up of companies who would be invited to tender. One noticeable omission was that of the name Shorts, indicating perhaps the P-17A/PD17 was no longer considered a viable proposition. Shorts had put in a considerable amount of work in an effort to reduce the number of engines PD17 was reliant on, but this work ceased and the

# EARLY DAYS

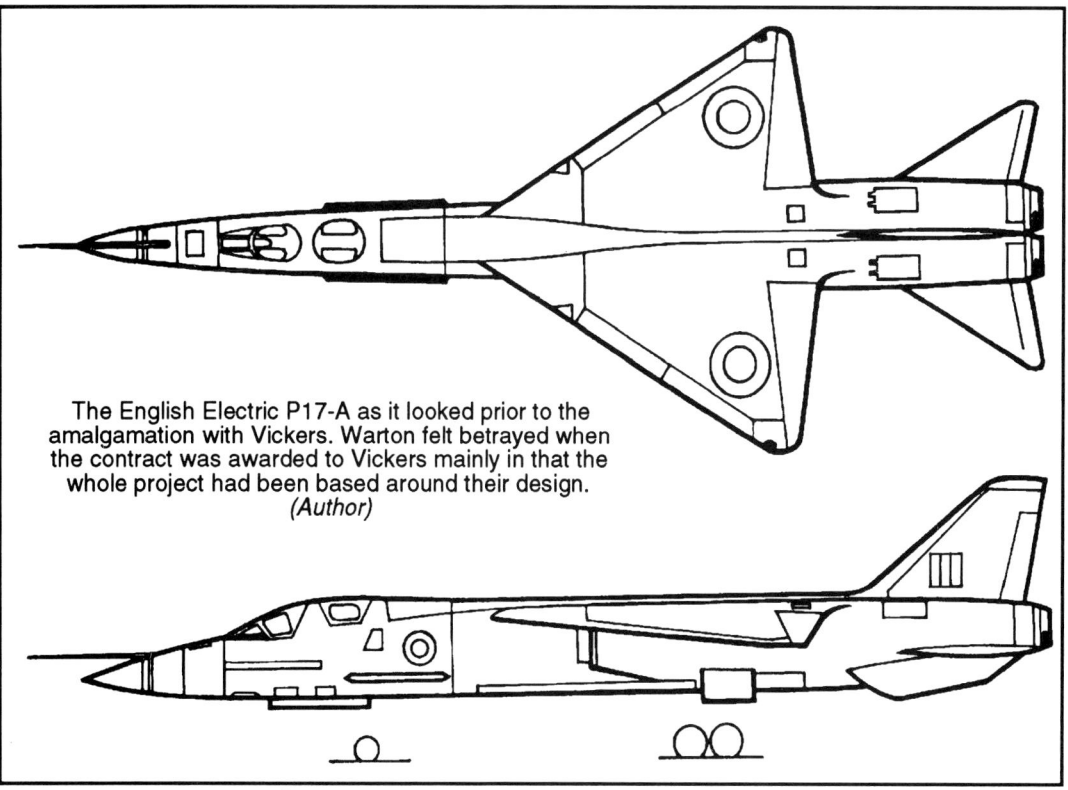

The English Electric P17-A as it looked prior to the amalgamation with Vickers. Warton felt betrayed when the contract was awarded to Vickers mainly in that the whole project had been based around their design.
*(Author)*

Preliminary Design 17 project never got beyond the drawing board. Of particular concern was that Shorts were no longer being considered with their 'lifting platform' proposal and this effectively left English Electric without an advantage as the merits of P-17A were virtually in line with its contenders.

During 1958, English Electric continued developing the P-17 and had submitted, a quarter scale model of the engine intake to both Rolls-Royce and Bristol Aero Engines, so that the engine manufacturers could assess the design and foresee any inherent problems that would affect the performance of their engine, should it be chosen. After the first draft of OR339 had been issued, a number of detailed modifications were made and in August the second draft was issued and later further modifications required a third draft in November. It was in that same November that English Electric and Vickers began what was to be a more formal meeting, where the technicalities of the project were to be discussed. Although up until now Vickers had made only tentative design studies, thus far there had been no attempt to consolidate the design from the two companies. English Electric were at an advantage having instituted a proposal for such an aircraft together with the MoS, so their design was at a more advanced stage. Vickers on the other hand had only made vague, basic studies. A close look at their original Type 571 proposal will give some idea as to how the company was thinking.

Their original submission to GOR339 was a single engine study but had now developed into a twin engine design. These were side-by-side in a thin fuselage, similar to that of the P-17A, but shorter and had engine intakes well forward. It had a swept shoulder mounted wing with leading edge slats with slotted flaps and carried wing tanks. The aircraft was heavier than P-17A but its wing area was smaller,

## TSR-2 PHOENIX OR FOLLY?

giving it a higher wing loading. The tailplanes and fin were all moving units with the tailplanes having a pronounced anhedral, and there was also a braking parachute stowed in the fairing at the base of the fin. The undercarriage, although conventional in appearance, retracted forward.

**From the Ashes of the Old.**
There was now growing evidence to show that Sandys' thinking was even at this early stage, beginning to lose some credibility. Although the momentum of early cancellations had already taken their toll, there was a realisation that not all conflicts would require a nuclear deterrent and that a small crisis might only involve the sending of a few aircraft to demonstrate the need for constraint. This alone illustrated the requirement for conventional fighters and bombers and the need to continue with P-17A became even more important. By December 1958, further modifications necessitated a fourth draft of the OR and it was during this month the whole TSR-2 project became public knowledge when a question in the Commons required the Secretary of State for Air to reply to a question from Geoffrey de Freitas. It was also announced that the newly-formed Bristol Siddeley Aero Engines Ltd, were to supply the powerplants for the new aircraft. The MoS formally issued a Parliamentary statement on January 1st, 1959, to the effect that a contract had been awarded to Vickers-Armstrong, (Weybridge) and English Electric, (Preston) for a new aircraft, to be known as TSR-2. (Tactical Strike and Reconnaissance - Mach 2, it was originally to be known as Tactical Support and Reconnaissance). Therefore in 1960 three companies officially formed an association known as the British Aircraft Corporation, (BAC), with Vickers and English Electric having an equal 40% share and the Bristol Aeroplane Company 20%. What was the Hunting Percival Aircraft Company was to become part of the group which, in 1977 was to be known as British Aerospace, (BAe).

It was Lord Knollys, Chairman of Vickers, who was to suggest that BAC's first Chairman should be Lord Portal of Hungerford. Portal was a man who knew nothing of the aviation industry, but had been involved in aviation of one kind or another for much of his life. In 1940 he had been Chief of the Air Staff and had served under Churchill. Portal was not a young man, but even at 67 years of age, he was to bring his strong character to BAC that would weld the companies that formed it together. Sir George Edwards, of Vickers, was to become Deputy Chairman and later Managing Director, and was to be Portal's right hand-man. This was a combination that was to work very well, with both men having great respect for each other. Edwards involvement in aviation stemmed from the days in the design office, and the Vickers Viscount had been one of his greatest achievements. It was suggested that Portal merely became the launching pad for Edwards ideas, but those who knew them or were connected with BAC, knew very well that this was not the case, but simply demonstrated how well the two men worked together.

Portals great mind had quickly grasped the concept of TSR-2 and he could see that this would be the aircraft, not only to establish BAC as a major world aircraft manufacturer, but to put Britain in the lead in aviation technology, giving the Royal Air Force the most advanced aircraft in the world in military service. English Electric's Chairman, Lord Nelson,

Lord Portal of Hungerford arrives at Filton for a meeting with Reginald Verdan-Smith. Portal was BAC's first Chairman and with the help of George Edwards, was to lead the company through its formulative years. *(BAe)*

felt his company had somehow been betrayed, for on a number of occasions English Electric had demonstrated how viable the company was with its high workload and sound financial standing. In the past English Electric had concentrated on military projects and knew well the procedures and problems at Whitehall. For a number of reasons, English Electric's standing at Whitehall appears to have been badly dented, because it was Vickers who had been appointed as the prime contractor, even though the company appeared to have a number of problems that were affecting its workload, and also low sales figures which were giving rise to other problems. The bitterest pill however, was the fact that this whole project was based round Warton's P-17A design, the most important ingredient in the new project.

**Amalgamation.**
An official design team was formed at Weybridge in February 1959 and both English Electric and Vickers worked together on the design specification. But it was not until early April that a design study contract was made between Vickers and English Electric. This amalgamation was to bring many problems to all the separate manufacturers involved in the forming of this new company. At the time each manufacturer was deeply involved with their own projects. Vickers still had a number of Viscounts to build, and the Vanguard was being held up due to problems with the Rolls-Royce Tyne engine. The VC10 had not yet flown and a super VC10 was in the pipeline. English Electric on the other hand were involved with all-military equipment. Its main production of the Canberra was slowly coming to an end with over 1400 of these aircraft being built, including those built in America and Australia. The main production was to be the supersonic Lightning. Some 25 aircraft were now flying and orders for Mk.1, Mk.1A, Mk.2 and the two seater Mk.4 had been received. Hunting Aircraft, one of the smaller manufacturers, was deeply involved with the Jet Provost, the RAFs jet trainer. Bristol had by now virtually completed the production run of the Britannia and were doing a feasibility study of a new supersonic airliner that was to carry 130 passengers at Mach 2.2 over 3,000 miles.

In an attempt to overcome the many problems, it was decided to place each of the companies projects into a new accounting system, 'New-projects' and 'Old-projects'. This meant the Old-accounts' remained the overall responsibility, both financially and contractually, of the originating company. Workers would continue on the projects as before, with the financial responsibility being looked after by the accountants of the companies involved. The 'New- accounts' for English Electric were made up of the Lightning Mk. 5 and onwards, Thunderbird II and Blue Water (both missile systems), and TSR-2. Bristols new accounts were the Belfast wings, Bloodhound 2 and a Supersonic transport, TSR-2, plus various other research and development contracts. Vickers had VC10, VC11 and TSR-2. This changeover was to take some four years and it was during this transitional period that all the companies involved, not only had problems of current production and development, but also of designing and building this completely new and controversial aeroplane.

**Further Demands OR.343.**
Because of the complexity of TSR-2, a far more exacting Operational Requirement was issued in addition to OR339, and known as OR343. The Air Staff, in conjunction with help from Vickers and English Electric, began a detailed specification for TSR-2, a demand that was to put further pressure on the

manufacturers who were still having difficulties with the specification in OR339. More demands were made from the engine, an area already in trouble even at this early stage, and in view of this a review of the materials used in the manufacture of the airframe was made in an attempt to contain the performance figures demanded in the new OR343. The first draft of OR343 was issued in May 1959 and some of the more notable points are listed below.

- The aircraft must be able to attack targets up to an extended range of 1,500 miles.
- Fly the specified portion of this range at very low level and high subsonic speed.
- Attain Mach 2.0 at the tropopause when required.
- Deliver nuclear and/or conventional weapons from low and medium altitudes in poor visibility and at night.
- Carry out an all-weather photo reconnaissance.
- Operate from small semi prepared fields with restricted maintenance facilities.
- Require no defence armament.
- Primary emphasis on low-altitude penetration, particularly in Europe where defences were arrayed in depth.
- Provide best gust response characteristics consistent with airfield performance to ensure maximum operating efficiency of crew.
- Carry maximum photographic and/or radar equipment for reconnaissance without prejudicing the strike role.
- Remain serviceable in the open with no service support for three days and with only minimum support for thirty days.
- Be weatherproof in flight and suitable for global operation.
- Have sufficient forward and down view to enable control of aircraft during all flight phases.
- Windscreen must be capable of withstanding a 3lb bird strike at 750 knots and be kept free of rain and insect debris at all times.
- An escape and survival capability is essential over the entire flight envelope of the aircraft.

By now a study contract had been issued to cover work between January and December 1959 and in October English Electric designers had returned to Warton. Although a draft contract was issued in August 1960 to Vickers for the production of 9 development aircraft, a firm contract was not issued until October. Delays in the issuing of contracts were to occur on a number of occasions and eventually were a contributing factor in the project slipping behind its schedule. This also became a problem that would affect the relationship between Whitehall and BAC.

**Dictating the Build.**
Although the overall design of an aircraft comes from the drawing board and is influenced by the equipment the plane will have to carry, the Treasury in holding the purse strings, can, through the Ministry, dictate what and who, supplies that equipment. In TSR-2 there were three categories and this was to have an effect on the final outcome of TSR-2. Category One, was for equipment specified by the Ministry and its suppliers. Equipment involved was the inertial navigation platform, the forward looking radar and the reconnaissance system. The engines also were to come under this category and as we shall see later under the Olympus story, this was to cause some concern, especially in the later development of the

# EARLY DAYS

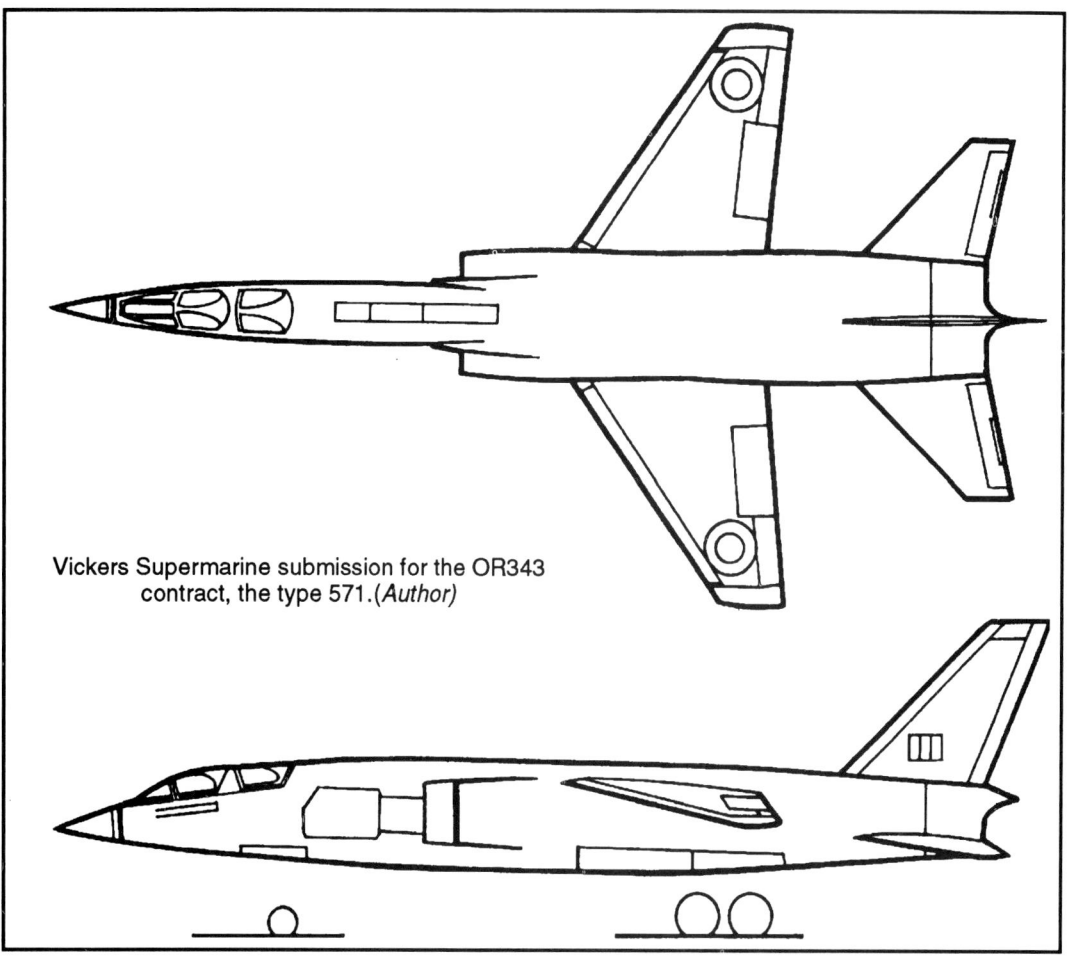

Vickers Supermarine submission for the OR343 contract, the type 571. (*Author*)

engine. Category Two, was Ministry authorised equipment but purchased by BAC, and included the central digital computing system, the sideways looking radar and the automatic flight control system. Category Three was equipment bought by BAC not needing Ministry authorisation and this included equipment such as wheels, tyres and the hydraulic system.

**A plethora of meetings.**
Great rivalry existed between Weybridge and Warton; Vickers had been awarded the contract, but it was felt this was a Warton aeroplane, as the contract was based around English Electrics P-17A. Rivalry at times became hostile and one can only admire the dedication of the engineers from both sites who were responsible for the central section where the mating up of the forward and rear sections took place. When the two halves came together, in March 1963, very few problems were to hold the project up thanks to the commitment of these few men.

In an effort to ensure mistakes made in the past, such as those with the Supermarine Swift were not repeated, an attempt was made to implement a form of management control. The Ministry was to hand control to a single contractor who would then sub-contract work out to other manufacturers. This was agreed by the Minister, Aubrey Jones, who also recommended the 'Weapons System' method be imposed originating from the 1955 White Paper. To eradicate many of the problems the new production methods were likely to bring a number of committees were formed. In the majority of cases the Ministry appeared to be the influencing factor and sent invitations to some of the main contractors, mainly BAC.

Two committees were formed to deal with the financial aspects of the TSR-2 project, one being chaired by an assistant secretary from the Ministry with BAC having representation. The second committee had Treasury influence, but there was no representative from BAC or any other contractor. Soon after the major TSR-2 contract was awarded, the various 'Design Authorities' were nominated formally for the specialised areas of the programme, among then Dr Henry Gardner, Director in Charge and Roland Beamont, chief project test pilot. The Systems Integration Panel was formed in 1960 and met every two months to deal with TSR-2's complex navigation, and flight systems. Beneath this, running in parallel were the individual sub-panels, responsible for Nav Attack, Flight Control Reconnaissance and also the Weapons Carriage and Release Systems. All these sub-panels were chaired by the relevant 'equipment director' and attended by some twelve members. The TSR-2 Management Board was also formed in 1960 attended by RAF officials and representatives of the firms. This committee convened every month during the early days but was later reduced to every quarter. The RAF Projects Policy Committee covered the HS681, P1154 and TSR-2 projects and was attended by similar officials who would in turn invite representatives from the various manufacturers. In 1964 this group was merged with the TSR-2 Steering Committee which had been founded in 1963. This particular committee was senior to all of the various committees and was chaired by the Minister of Aviation himself, with George Edwards on the panel representing the manufacturer.

The original intention had been for these committees to ensure continuity of work and at the same time attempt to control costs. This was made all the harder by sub-contractors approaching the Ministry direct when seeking assurances regarding costs or specifications of components for which they were responsible, instead of BAC. This effectively took the responsibilty and control from BAC's hands, leaving them unable to curb the increasing costs. The committees themselves, although specific in their tasks, had no direct power with which to enforce control and decisions made invariably had to have Ministry approval. The validity of the decisions made by the committees was also questionable, and was amply demonstrated in an instance quoted by Roland Beamont in one of his books. Apparently one committee, consisting of some 40 people, from the manufacturers and the Ministry, took nearly a day just to decide on the location and position of one switch and its associated label! One cannot help thinking that a decision of this scale and nature could have been made by the men who were to test fly the aircraft, probably in less than the time it took to assemble the above mentioned committee! The numbers attending the committees also came into question and again from Roland Beamont's publications, reference is made to one particular committee where the Chairman asked for a head count. It was found some 55 members were at the meeting, so the Chair requested this number be reduced for the next meeting. When they next convened it was found the number had increased to 60!

**Type 571.**
After the discussion by English Electric and Vickers there appeared a better understanding of the technical needs of the new aircraft and by January 1959 both companies were preparing detailed design work on the components each was responsible for. Vickers were to be responsible for the front half including the

# EARLY DAYS

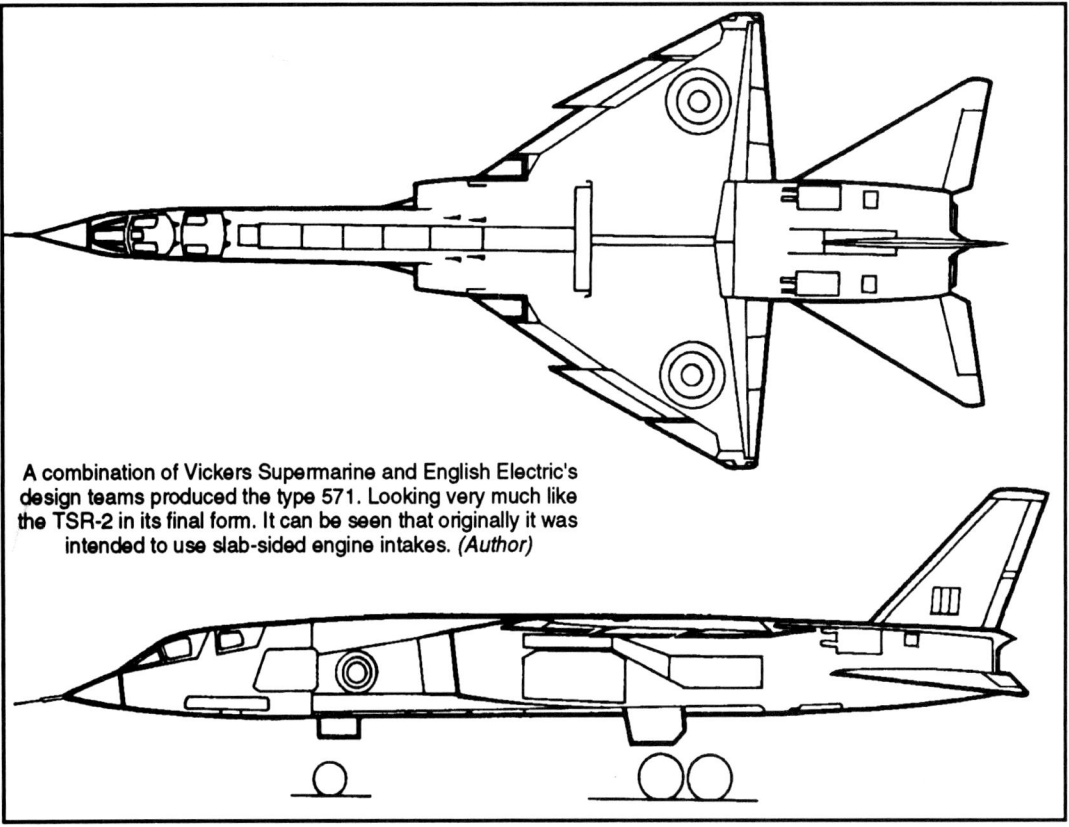

A combination of Vickers Supermarine and English Electric's design teams produced the type 571. Looking very much like the TSR-2 in its final form. It can be seen that originally it was intended to use slab-sided engine intakes. *(Author)*

cockpit and its layout, electronics and armament installation, whilst the English Electric plants were to be responsible for the rear fuselage, wings, engine installation, fuel system, powered flying control and autostabilizers. By July 1959 most of the intricate design work had been completed and it was decided the two companies should issue their brochure. Vickers Supermarine and English Electric had agreed the prototype batch of aircraft should adopt the Vickers designation, Type 571, and reference to P-17A was to be dropped.

The VS/EE 571 was an aircraft very much like TSR-2 in its eventual form and was to change only very slightly. The fuselage was long with slab sides up to the engine intakes, and this forward part would contain all the sideways-looking radar, ground terrain following radar, avionics and systems equipment. Behind this equipment in the systems bay the No 1 fuel cell was located. Forward of the avionics bay sat the pilot and navigator in tandem, sealed in separate pressurised cockpits with individual escape systems under clam shell type cockpit canopies. Each side of the fuselage and slightly forward of the wings leading edge were the hydraulically controlled slab type engine intake doors. The large delta main wing had a stepped leading edge with the outer half forward of the inner half and had high lift mechanisms to assist with aircraft stability at low speed. The trailing edge of the wing although straight, swept very slightly forward and contained large, powerful, flaps. The wing itself contained very few ribs except were it was intended to mount external stores or wing-tanks, with the majority of the wing being a huge fuel tank. The fuselage at the point where the engines were accomodated had become almost square. The engines were mounted in separate tunnels, with fuel cells just to the

# TSR-2 PHOENIX OR FOLLY?

fore and water tanks between the engines. The weapons bay ran the majority of the length of the fuselage with the undercarriage legs on either side that would retract forward and be stowed in wells within the fuselage. The aft section, apart from containing the Olympus engines, had three large air-braking devices, two on the top and one in the centre underneath. Just to the rear and mounted half way up the fuselage, were the one piece, all moving, tailerons with a 4° anhedral. At this time it was still undecided whether the fin should be all moving or not. The engine exhausts exited just below the rear fairing containing the braking parachute.

**Metal is Cut.**
By 1962 there was an urgent need to commit the design to metal, as additions and amendments were being continually submitted. Therefore, it was decided to freeze the present general arrangement in order that building the aircraft could begin. The project was now some 18 months behind schedule, and, with the engine problems, there was a danger this delay could get worse. Still designated type 571 and following closely the design described earlier, metal was cut. The biggest external differences were in the main wing, fin and engine intakes. After wind tunnel tests on the wing it was found the need for high lift devices on the leading edge were unnecessary and so the wing lost its 'saw-tooth' leading edge.

TSR-2 XR219 ready for the paintshop at Weybridge (BAe)

It was found that introducing a distinct anhedral to the wing-tips would increase the lateral stability at low level and low speeds. Without this unusual design such stability would have been greatly reduced by the original proposed 'straight' non-anhedral, lightly swept wing. The trailing edge, however, retained the blown flaps.

The designers had settled on an all moving fin and had slightly increased the overall area. The engine intakes had changed from the slab type to hydraulically controlled half cones. Although the two air-braking devices were retained on the top of the fuselage, the single, ventral device was removed and two air-brakes were also fitted underneath. Other, less noticeable, changes were in the avionics and in small design changes within the fuselage. This was built in frames and was calculated in imperial inches, starting from the nose of the aircraft. Because of a slight change on the drawing board, instead of starting at frame number 1, it actually started at frame number 4. The two halves of the fuselage were mated up between frames 629 and 640. TSR-2 itself is looked at in detail during the next chapter.

The TSR-2 was very much an aircraft designed with advanced thinking and to ensure that it would be capable of carrying out these specific roles right into the 1970's and 80's, the RAF had tried to include every possible system they believed would be needed during the years it would be in service. It was a complete weapons system and could be truly called a 'Multi Role' aircraft. When first envisaged its role was that of a Tactical bomber with reconnaissance capability, and the fact that it could also be used as a Strategic weapon was a bonus and one that caused a certain amount of controversy later on in its development.

# CHAPTER TWO
## TSR-2 The Plane.

Today it is generally accepted that the modern military aircraft is perhaps one of the most complex weapons of war. To support this view, it could be said that the equipment carried by one of today's aircraft, such as the Tornado, would very possibly in the 1960's, for example, have been considered so futuristic in concept, as to be almost unreal. Therefore TSR-2 itself could easily be described, as the most exacting project undertaken by the British aircraft industry up to that time. The huge amount of research and development programmes initiated by the building of TSR-2 were without comparison and involved many industries not generally associated with aircraft manufacturing, with the electronics industry reaping the most benefits.

The many new and complex working methods introduced into the project quite early on, were to alter significantly those methods already being used by the industry for the manufacture of prototype and production aircraft. Adopting such new methods meant that not only did BAC have to change some of its normal working practices with respect to its current projects, but it also had to define a method by which these two different practices could be integrated within the company organisation itself. This, no doubt not only increased the pressure on the two company bases at Warton and Weybridge, but also added to the pressures that the project was coming under. Although the eventual production of the TSR-2 aircraft would be at Salmesbury, the building of the development batch was focussed at Weybridge. Whilst the methods of construction and the ensuing consequences are studied later on, a brief outline of the basic problems will give an insight into the common causes of dispute.

*The TSR-2 production line at Weybridge. Building methods previously used by the industry were to alter significantly with the TSR-2 production. (BAe)*

# TSR-2 PHOENIX OR FOLLY?

**Early Developments.**
In Britain a relatively simple method was used to develop and produce military aircraft in particular. Usually the manufacturer would hand build a single prototype aircraft. This would allow him to prove its basic concept whilst at the same time enable him to forsee any problems that were likely to affect a future production run. English Electric had used this method to develop and produce the Lightning.

The Lightning was conceived from the P1, an experimental research programme and supersonic fighter requirement, F23/49. Designed to explore the problems associated with a supersonic fighter, two P1's were built, and the experience gained from building these aircraft provided English Electric with the knowledge they needed to develop and build the P1-B Lightning.

The development and production of the TSR-2 was to alter significantly in that the programme was to adopt an American method of build and control. An assembly line would be built using production jigs and tooling, thus the prototype aircraft would be built in production jigs. Whilst this American concept did have advantages, the biggest argument against this method was that it delayed the programme considerably. Perhaps the biggest departure was the fact that the airframe was to be built in two separate halves at companies so far apart. Although all military aircraft in Britain were manufactured in small sections, the sections themselves were built within the same company and invariably within a short distance of the main production line.

**Describing the Aircraft.**
Because of the many varied duties TSR-2 was required to perform, the aircraft was specifically designed to achieve an optimum performance as a tactical strike and reconnaissance vehicle which would be able to perform not only at high speed at high level, but also with a transonic, low-level, terrain-following capability. This was achieved by designing an aircraft with a small delta wing that gave a high wing loading but with a low gust response. The fuselage was 89ft in length and carried a crew of two, the avionics bay, fuel tanks and at the rear the Olympus engines.

**The Fuselage.**
The manufacture of the long fuselage involved the use of some new materials to provide the airframe with increased ability to withstand the varying temperature and pressure changes it would face as a multi role aircraft, and also to ensure that the high fatigue life of something like 3,000 hours in the airframe could be achieved. It

This air-to-air shot ideally shows off the long fuselage that was to hold the avionics equipment, most of the aircraft's fuel and at the rear, the twin Olympus engine's. *(BAe)*

# TSR-2: THE PLANE

was anticipated that the fuselage would be subjected to high temperatures due to kinetic energy, especially affecting the middle and rear sections of the aircraft, and therefore these areas were manufactured in aluminium-lithium. The most extreme temperatures were expected to be centred around the aircraft's twin jet exhausts, particularly during any extended use of reheat. Accordingly, this area was to be manufactured in Waspalloy. This rearmost section of the tail, a piece always seen in its natural finish, was the responsiblity of Bristol Siddeley and was manufactured at Filton. Other parts of the airframe, not subject to such high temperature changes, were manufactured in aluminium-copper alloy.

The actual manufacturing techniques used in the construction of the TSR-2 fuselage and wing had been pioneered by Vickers during construction of their Vanguard and Viscount passenger airliners. This technique relied on the integral construction using stiffened skins, machined from stretched billets and also included the use of chemically etched skins. Although the aircraft's skin was considerably thinner than that used in the construction of the airliners, the desired stiffness was achieved by an integral construction of machined deep flanged stringers.

The comfort of the crew had been an important consideration, particularly when it was realised that TSR-2 would be flying at around Mach 0.9 at only 200ft from the earth's surface. Many of these sorties would be flown in extreme bad weather conditions, subjecting the aircraft structure to severe buffeting and vibration putting the crew under considerable strain. To provide some relief, the construction of the long fuselage, specifically in the area of the cockpit, had been specially designed to flex, thus alleviating the fatiguing effects of constant buffeting. This particular problem had been highlighted by the pilots of the chase aircraft who when trying to keep low down with TSR-2, had been forced to break off and seek the calmer air higher up leaving TSR-2 to continue flying along at low level, with the crew unaffected by buffeting.

**Protecting the Crew.**
Flying at such low altitudes also posed a number of problems, one in particular was the threat of bird strikes. This presented a greater risk than normal for TSR-2 flying at these altitudes and across differing types of terrain. Therefore the protection of the crew, the pilot in particular, was extremely important. To this end BAC gave the job of developing and manufacturing suitable window protection to the Triplex Safety Glass Company. The design and manufacture of the windscreen had to consider a number of problems. Firstly the threat of bird strikes dictated the screen had to be strong and able to withstand the impact of a bird approximately 3lbs in weight. Secondly, TSR-2 was to use a hot-air-rain-dispersal system. This utilised hot air bled off the engine which was then blown across the windscreen to disperse insect debris and water in adverse weather conditions. Also as a result of the aircraft's supersonic role, heat externally generated would have to be a consideration. Thirdly, the Head-Up-Display was to be projected directly onto the windscreen which neccessitated the use of an exceptionally high quality optical glass. There was of course the need to be able to demist not only the windscreen, but also demist and de-ice the quarter light windows as well.

Triplex developed and manufactured a multi-layer screen with the outer layer using a low expansion glass type, then an interlayer of Silastomer K8 to help give the screen strength. Silastomer was able to withstand temperatures up to around 150° C. The next layer was perhaps one of the most important in that it was the optical glass. On its outer side was a film of gold which together with elements embedded in the glass were used to give quick and even demisting over the whole screen. The inner side of this glass had a coating of reflective material onto which would be displayed the HUD information. The quarter windows used similar techniques to demist the glass, also the elements within the glass were used to de-ice the glass when necessary. The rest of the crew canopies were manufactured in Plexiglass 55, using three layers to provide strength and used the same method for demisting and de-icing as that used on the quarter light windows.

On XR219 and XR220 in particular, it was found that the plexiglass transparencies had an unacceptable amount of wave distortion and would require a modification for future types. Comment has also been made by some on the effectiveness of the HUD being displayed directly onto the windscreen. The TSR-2 screen was very big and in the opinion of those who worked on this particular aspect of the project, it was felt that the amount of distortion on the screen would have made the information unreadable under certain conditions. The American F-14 Tomcat was specified with a similar display, but had to resort to a conventional display whereby it was directed onto a secondary plate as in all other aircraft with a head-up-display.

**Crew Comfort.**
Access to the individual cockpits was by ladders on the starboard side clipped to hardpoints in the fuselage. The navigator's compartment housed a folding ladder for use when TSR-2 was away at an unsupported field. The cockpits themselves were particularly spacious and, like the seating, had been designed with crew comfort in mind. The actual seats were manufactured by the Martin Baker Co, to a BAC specification, and as endorsed by the pilot and navigators, were unusually comfortable for a military aircraft. They were Mk.8VA, rocket assisted ejector seats, which could quite easily be adjusted to suit the individual crew needs. The whole seat assembly was fully adjustable with the pilot able to raise or lower the seat and move it forward or back electrically. This was particularly useful during the landing phase or on a bombing run when he needed a clear view over the nose of the aircraft. Martin Baker had designed this seat to ensure the crew could safely eject from zero-zero circumstances up to 56,000ft and at speeds up to 650 knots IAS or Mach 2.0, whichever was the least. The escape system was designed so that either crew member could initiate ejection. However, should the pilot initiate first, it would automatically eject the navigator. The navigator being responsible for initiating only his own ejection, a system used on Tornado today. There was the risk that the navigator could be unprepared for a sudden, unannounced ejection, either because he was injured, or there was no time to inform him, particularly if the aircraft was on a low-level sortie. To avoid this a method was devised whereby the navigator could be safely ejected without him sustaining injury to his back, legs or arms, when leaving the aircraft. Built into the seat were leg and arm restraints, which meant that when, irrespective of the position of the crew member, either ejection handle was pulled the restraint automatically pulled the legs into the seat-pan and the arms into the seat. Sufficient slackness had to be left in the arm restraints to allow the crewman to initiate an Emergency Harness Release in case

The Martin Baker MK 8VA rocket powered seat in TSR-2 was, according to those who flew the aircraft, one of the most comfortable seats in a military aircraft. *(Author)*

## TSR-2: THE PLANE

the automatic sequence failed. The shoulder harness too was pulled tight automatically by a ballistic motor operated by gas pressure. This also pulled the head restraint in, so by the time the seat fired, the crewmans body position was such that when he went out, his arms, legs or head would not flay about risking injury or even amputation. There was also a system, whereby the crew, wearing a specially designed helmet, would have the face visor pulled down automatically. So in ejecting from the TSR-2 the crew would be correctly positioned within the seats so as to eliminate injury. This procedure would take only a matter of seconds.

The final part of the escape sequence was the charge for the ejection of the crash recorder equipment. This equipment, manufactured by Redifon Ltd, was housed in a unit within the fuselage and could be ejected under a number of circumstances, particular examples being when the pilot ejected, when a preset altitudinal deceleration was triggered, when subjected to pressures in excess of 23lbs./sq in, (i.e should the aircraft become submerged in water deeper than 20ft), or when a sudden excessive rise in temperature was experienced. The recorder itself was ejected through a small breakable panel in the skin of the fuselage and had a small parachute which helped to decelerate its fall.

*This aerial shot of XR219 shows to good effect the main wing of the TSR-2. The wing itself was attached to the fuselage by 16 non-rigid bolts which allowed the wing to flex. (BAe)*

### The Wing.

The 60° delta wing with its 37ft wingspan was made up of machined aluminium planks and was secured to the fuselage by sixteen non-rigid swing joints. This allowed the wing to flex on its mountings. During the early development stage it had been suggested the aircraft should adopt a swing-wing capability, similar to that which was being designed for use on the the TFX, (F-111). Although such a wing may have given TSR-2 more flexibility, it was considered that a device such as this would be inappropriate, due not only to the increased weight, but mainly because of the excessive costs that would ultimately be involved in its development.

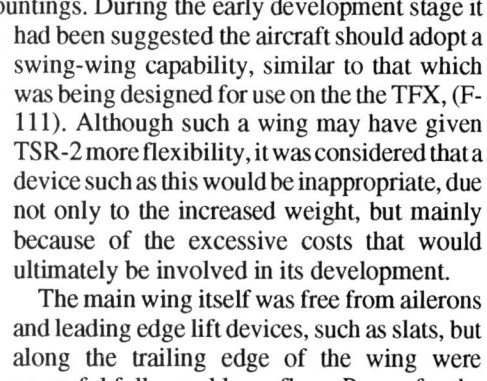

The main wing itself was free from ailerons and leading edge lift devices, such as slats, but along the trailing edge of the wing were powerful full span blown flaps. Power for the flaps was bled from the high-pressure compressor on the Olympus engines and, by using an intricate cross-over system, allowed one engine to supply air for both flaps in the case of a single engine failure. The wing itself contained nearly one third of the aircraft's fuel load, and was capable of carrying external slipper tanks and stores on four hard points.

As the main wing lacked any anhedral or dihedral the wing tips had a 37° dihedral section to restore stability at low speed/low-level flight. This would otherwise have been provided by 2° dihedral on the whole wing. Also contained within the wing tips were various aerials and the wing tank fuel vents.

## Flight Control.

The hydraulically controlled flying surfaces on a normal aircraft are usually found on the wings as ailerons, the tailplane as the elevators and fin as the rudder. The large, full span, blown flaps on TSR-2 left no room for ailerons, and therefore roll control was by a differential tailplane which performed the function of elevator and ailerons. Each of the 'tailerons' was a large one piece unit which could move either independently to provide roll or together to give pitch. They were each operated by a tandem piston jack, with each jack operated by pressure from the Control Services hydraulic system. Should one system fail, the jack could still operate though at a somewhat slower rate. The jack control valve could be operated either by a mechanical linkage or by a duplicated Automatic Flight Control System actuator. The movement of the taileron varied from 10° nose up to 20° nose down, with the level flight setting at a 6° nose down angle. An artificial feel was built into the system to produce a constant feel in the stick throughout the whole flight envelope. With the roll force gearing, it meant the stick force necessary to achieve a given manoeuvre remained constant under all conditions. The reduction in taileron movement rather than an increase in stick feel ensured that the pilot did not create an excessive roll rate at high speed by exerting undue pressure on the stick. The tailerons were also fitted with follow up flaps to provide adequate response at slow speeds. This flap trim control was operated in conjunction with the main flaps. When the the main flaps were raised a hydraulic stop was initiated locking the taileron trim-flaps in line with the taileron chord. The fin which provided yaw, was a single slab, rising some 24ft above ground level. This all moving unit was operated by a twin piston jack, again supplied by the Control Services hydraulics system. It was connected by a mechanical linkage to the rudder pedals and by triple automatic flying control systems actuators. Because of the size of the fin it was found that a movement of only +/-12.5° was needed. The fin had been deliberately designed undersized in order to enhance the fatigue life. For this reason there was a dependence on artificial stability under high speed conditions.

## The Main Landing Gear.

The undercarriage on TSR-2, manufactured by Electro-Hydraulics, was to become one of the major problem areas and in fact delayed the beginning of high speed flight trials. In a later chapter some of the problems experienced with the undercarriage in the flight test programme are described.

In order that the undercarriage could cope with semi prepared runways, various components of the gear were manufactured in nickel-chrome-molybdenum-vanadium for added strength. The gear itself consisted of two main bogies with large tandem wheels using low-pressure tyres. The oleos were particularly long stroke in order to cope with the rough field landing. Each oleo was designed to be soft at initial touchdown, and as the heavy aircraft settled down, so it became stiffer.

The siting of the undercarriage was most important as it had to be accomodated in an area that was already filled with fuel, electronic equipment and the large Olympus engines. Care had to be taken to ensure the aircraft's centre of gravity was maintained when the gear was retracted in its forward position, this was not helped by the undercarriage legs having to be splayed out in order to fulfil the maximum track requirement of 13ft 6ins. A complex method was devised to retract and store the undercarriage in the smallest area possible. After take-off the main legs naturally extended due to the lack of weight, the bogies then rotated forward through some ninety degrees. This put the bogie in-line with the main leg with the

## TSR-2: THE PLANE

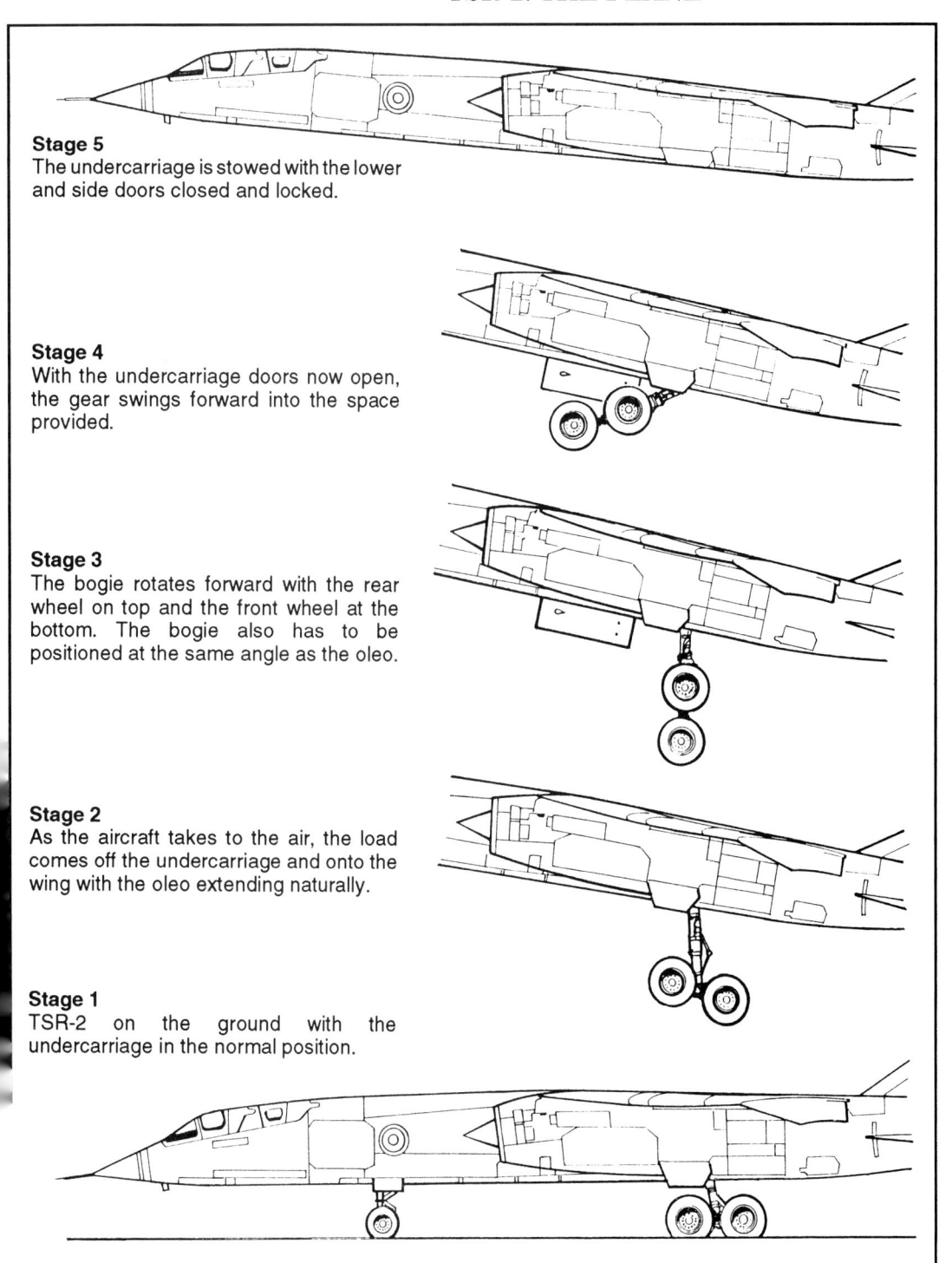

**Stage 5**
The undercarriage is stowed with the lower and side doors closed and locked.

**Stage 4**
With the undercarriage doors now open, the gear swings forward into the space provided.

**Stage 3**
The bogie rotates forward with the rear wheel on top and the front wheel at the bottom. The bogie also has to be positioned at the same angle as the oleo.

**Stage 2**
As the aircraft takes to the air, the load comes off the undercarriage and onto the wing with the oleo extending naturally.

**Stage 1**
TSR-2 on the ground with the undercarriage in the normal position.

The undercarriage sequence was quite complex, but was designed to allow the undercarriage to take up as little room as possible and not upset the centre of gravity when stowed. *(Phil Kingham)*

## TSR-2 PHOENIX OR FOLLY?

*Looking extremely robust, the undercarriage was designed to cope with semi prepared runways. Shown here is the modified operational strut fitted to XR220. It was also fitted to XR219 but was never proven as the project was cancelled prior to it being fully completed. (Author)*

front wheel on the bottom and the rear wheel at the top. The bogie now had to be angled in its vertical position to the same degree at which the legs splayed out. This would place the rear wheel snug into the oleo with the whole gear ready to swing forward into the space provided.

With the aircraft on the ground only a single door remained open for the main leg, other doors remaining locked until either the undercarriage sequence was operating or an 'emergency down' was initiated. In the retraction sequence the large side door opened and the lower door swung down to accept the undercarriage in the well provided. All doors then closed, each one having a micro-switch to indicate the door had closed successfully. Because of the complicated undercarriage sequencing, the main leg bogies had switches fitted to the bogie ankles indicating that the bogie had rotated and locked in the landing position. Each undercarriage leg was operated by a separate hydraulic system.

The large low-pressure tyres were a modified Canberra type, manufactured by either Dunlop or Goodyear. Dunlop also manufactured the disc brakes, each having a Maxaret anti-skid unit. These were fitted with electric cooling fans to cope with extreme braking conditions. The front and rear brakes were fed from two independent hydraulic systems to ensure even braking in case of a single hydraulic failure. In the case of a complete failure an accumulator was fitted to ensure a limited amount of pressure was available. Braking was activated by toe-operated pedals on the pilot's rudder pedals which could be operated together or independently to help with steering.

**The Nose Landing Gear.**

The nose wheel had a side-by-side wheel assembly, with the wheel steerable by the rudder pedals. The steering had a fine or coarse gearing system, which gave a fine steering angle with a maximum of +/- 10° or a coarse angle of +/-68°. A standby steering was automatically engaged in the case of an 'emergency down' selection, and when the steering was not engaged, the wheels castored between the fine and coarse steering limits and were self centring.

It was anticipated that during operations in the field with low operating weights, there would be occasions when the aircraft would require a short take-off. To assist with this the nose leg could be extended by some 30 inches. This would alter the aircraft wing angle to the ground which would eliminate the need for long take-off

*The nose wheel was designed so that the leg could be extended. A system that was used on the F4 Phantoms particularly for use from carriers. (Author)*

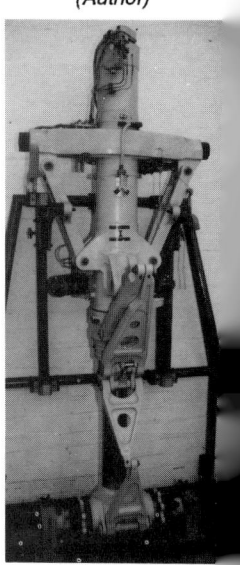

runs, particularly on semi prepared fields. The leg was extended by the pilot in the final stages of take off checks with the aircraft stationary on the runway. After take-off the undercarriage was retracted, with the noseleg contracting automatically. If the automatic operation failed or the pilot required the noseleg contracted he could do this manually. In the event that this manual operation failed there was an emergency facility to contract the leg but in using this facility would result in the loss of hydraulic fluid and was therefore discouraged. It was a development that was found not to be needed, as the power of TSR-2's engines and tailplane control were such that its take-off run was well within the parameters laid down in the Operational Requirement.

**Undercarriage Safety.**

Because of the undercarriage arrangement on TSR-2 some rather elaborate safety precautions were built into the system. This was to ensure that the large undercarriage doors were fully closed when the gear had been stowed. A facility was provided to accommodate the undercarriage failing to lower. The pilot would select an 'emergency down' switch. This operated two electrically driven change over valves which swapped the No's 1 and 2 hydraulic services over. The emergency pressure was taken in separate pipes direct to the release side of the equipment, hopefully freeing any unit that was still locked in the up position. Should this fail there was a last resort method that supplied a small cartridge which would 'blow' off the main door and leg up-locks. This system would not on its own lower any gear stuck in the up position and would require a hydraulic and electrical current to complete the phase.

On studying the detailed notes on TSR-2, there is no mention of a 'Bogie locked-up' situation, where the main leg has lowered but the bogie has not de-rotated to the landing position. It appears that although the pilot would have red lights to show the ankles were not locked in the landing position, then provided he had the necessary three greens showing the undercarriage legs were down and locked, he could continue with the landing, albeit very carefully. At the time there had been no briefing of this contingency and when Roland Beamont found himself in this position on flight 5, the problem was an unknown area.

**Hydraulics.**

TSR-2 had four completely independent, and self contained, hydraulic systems. These were divided equally to support the Control Systems and the General Services. Because of the widely differing operating conditions, TSR-2 was the first aircraft to use the DP.47 Silcodyne H hydraulic fluid exclusively for all its hydraulic systems. This fluid was capable of being used not only at high pressures, but also at high temperatures. Each system was maintained at a pressure of 4000lbs +/- 200lbs, with relief valves in line to ensure a pressure of 4800lbs was never exceeded.

**Control Services.**

It was necessary for the flying controls and essential services to have two hydraulic systems to ensure aircraft safety should one fail. Even in the event of a twin engine failure, both services had a back-up in the form of accumulators. These provided sufficient pressure in the system to allow for a limited amount of essential flying control movement. Although the amount of supply itself depended on the ambient conditions, it also depended upon the rate at which the pressure could be maintained if there was any windmilling action in the failed engines.

# TSR-2 PHOENIX OR FOLLY?

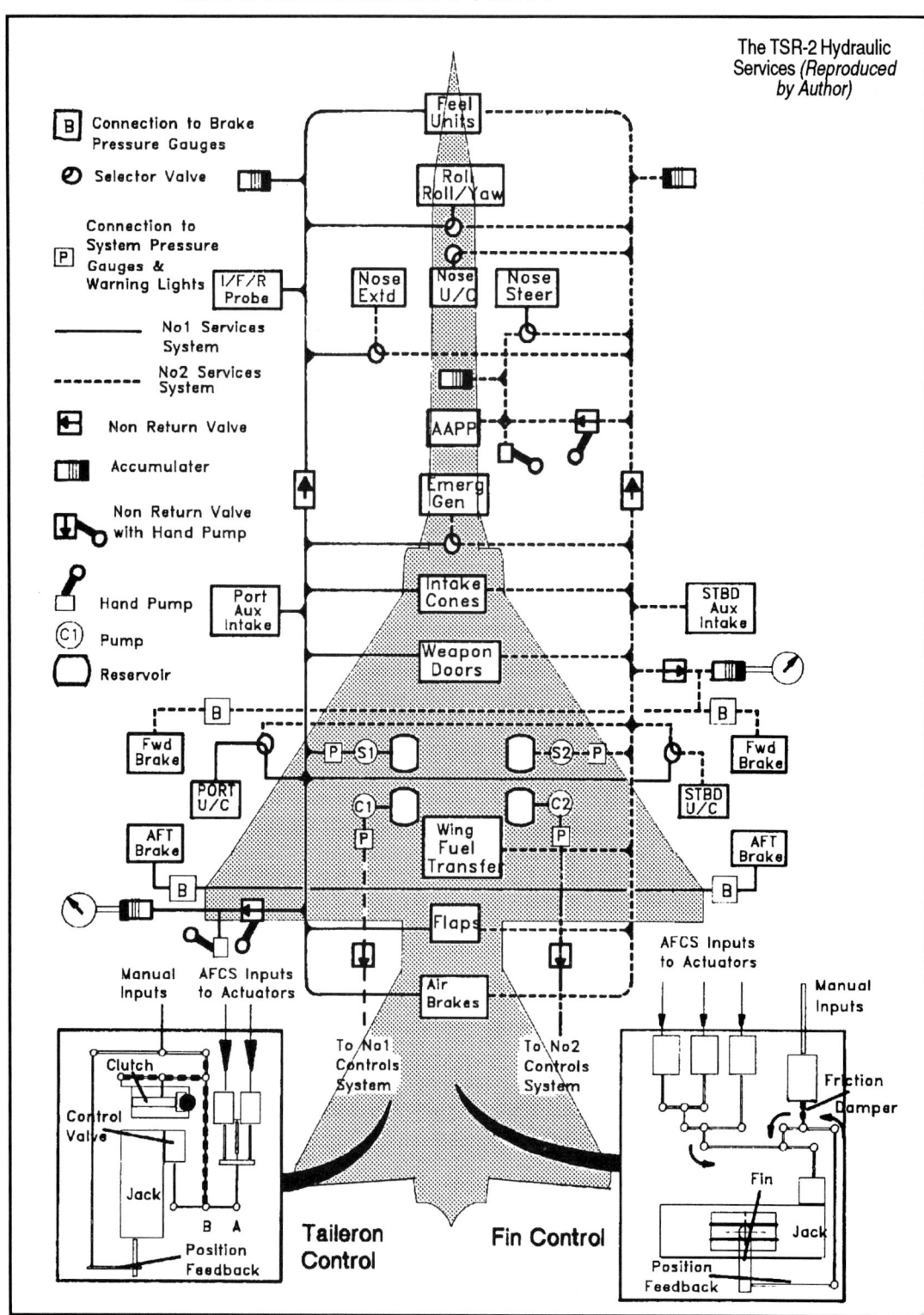

The TSR-2 Hydraulic Services *(Reproduced by Author)*

# TSR-2: THE PLANE

At altitudes below 35,000ft, the aircraft needed a forward speed of 220knots to maintain the vital pressure essential to allow the aircraft to be flown. The Control Systems accumulator also provided additional power during maximum rate manoeuvres.

**General Services.**
The General Services System had two services to maintain, numbers One and Two. Number One had two accumulators, one to provide pressure for the artificial feel unit and roll/roll yaw gearbox, and a second which supplied pressure for the rear wheel brakes on the tandem undercarriage bogie. The number two system had three accumulators. The first worked in conjunction with the number one services unit to supply pressure for the artificial feel unit and roll/roll yaw gearbox. The second supplied pressure for the front wheel brakes and the third supplied pressure for the Airborne Auxilary Power Plant, or A.A.P.P, together with emergency power for the nose steering in case the primary supply failed.

Gauges for the System services hydraulics were placed on the rear of the pilot's port console and displayed the output pressure of the system's pumps. Pressure to the two Service accumulators for the front and rear brakes was displayed on a twin-dial gauge situated on the port instrument panel. The navigator had four gauges to show the in-line pressure for each brake and these showed pressures ranging from 1750 to 1900lbs. All seven of the accumulators had an external nitrogen pressure gauge at various points around the airframe, and in addition to the various gauges externally and internally, the pilot also had four warning lights on his central display. The Control Services lamps displayed in red with the other services shown in amber. These were activated by a pressure switch on the appropriate system and would warn the pilot should the pressure drop below 2000lbs +0/-300lbs/sq. in.

**Blown Flaps.**
TSR-2 was the first aircraft to use full-span blown flaps. Although most aircraft use flaps, very few British aircraft of the time used an engine bleed system to provide air for the control of boundary layer air over the wings. Using flaps gives an aircraft lift, but in the slow speed ranges the air travelling over the wing becomes 'tired' and towards the trailing edge of the wing can become so sluggish that the extra lift created by the flap is overcome by the sluggish air above causing drag. By bleeding air from the jet engine, usually from the high compressor stage, it can then be vented out over the front of the wing, or in the case of TSR-2, out over the flaps at high speed; by doing this the tired sluggish air becomes re-energised, making the flaps more efficient. The Buccaneer used such a system for boundary layer control, which gave it the ability to fly some 25 knots slower during the landing phase than any similar aircraft not using a 'blowing' system. It was also particularly helpful during take-off from carrier decks, when this system gave more lift. The control of 'blowing' on the Buccaneer was across the inner and outer portions of the leading and trailing edges of the main wing, and there was also blowing across the leading edge of the Buccaneer's high tailplane. Although very efficient it was nowhere near as powerful as the blown flaps on the TSR-2.

The flaps on TSR-2 were controlled by three screw jacks in each flap. The jacks themselves were driven by screw shafts in the trailing edge of the wing connected to individual hydraulic motors. The hydraulic motors were served by a separate Control Services hydraulic system and in the event of this failing the secondary system would take over. The only adverse effect of this would be a slowing down in

# TSR-2 PHOENIX OR FOLLY?

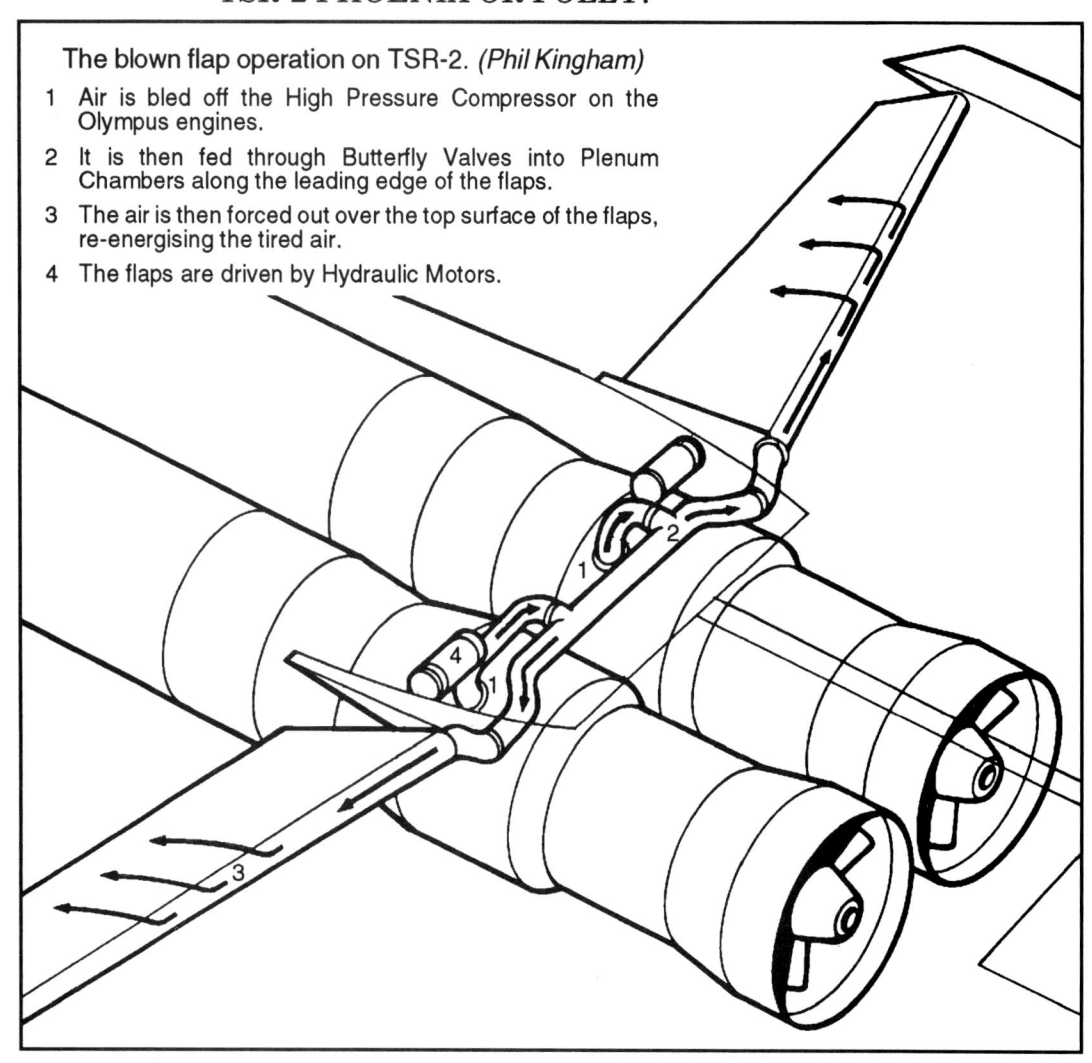

**The blown flap operation on TSR-2.** *(Phil Kingham)*

1. Air is bled off the High Pressure Compressor on the Olympus engines.
2. It is then fed through Butterfly Valves into Plenum Chambers along the leading edge of the flaps.
3. The air is then forced out over the top surface of the flaps, re-energising the tired air.
4. The flaps are driven by Hydraulic Motors.

the operation of the flaps. Control to the hydraulics was electrical with duplicate systems.

The flaps could be lowered to settings between 0 and 50° and there were indications at the 0, 20, 35 and 50 degree points. Flap blowing started when the flaps reached 15 degree and continued right through to 50 degree. The air was tapped from the high pressure compressor on the Olympus engine and fed through a butterfly valve for each system. The valves were controlled to ensure an optimum delivery of air was maintained throughout the flap range and to ensure the engine performance did not deteriorate during the blowing operation. The air was piped along the leading edge of the flap, then through holes into plenum chambers which ran the whole length of each flap duct, before finally passing through slots and out over the upper surfaces of the flaps. The system had a crossover feeding system to ensure blowing was maintained should an engine fail. This whole operation was fully automatic and required no pilot intervention, however the pilot could cancel the flap blowing completely by the operation of a switch in the cockpit.

# TSR-2: THE PLANE

Hunting Aircraft's H126 provided the industry with a great deal of knowledge in boundary layer control. This particular aircraft could fly as slow as 44 knots by venting the jet exhaust out over the aircraft's flaps. *(BAe)*

**Air Brakes.**

For speed reduction and increased rate of descent, TSR-2 had four huge air-brake doors mounted at the rear of the aircraft. Two were mounted on top of the fuselage, just to the rear of the wing trailing edge and two underneath. These brake doors were operated by screw jacks hydraulically driven by a motor, similar to the system used to operate the wing flaps. The brakes could be used with the primary or standby system. During early testing in the backup system a number of failures were encountered when the doors could only either be fully closed or fully opened. To ensure the air-brakes were not overstressed during flight, the system had a 'Blow-Back' control incorporated. This ensured that the screw-jacks were not overloaded by automatically closing the doors to a safe angle under extreme pressure. This blow-back system worked in primary or standby conditions.

**Fuel Systems.**

Besides the obvious complexities of the electronic systems, the next most demanding was the fuel system. Of course, this was the one that had to be sorted out completely before the aircraft was allowed to fly. The TSR-2 was capable of carrying nearly 52,000lbs of fuel and of this, a little over 44,000lbs could be carried internally. The main bulk of this internal load amounted to some 33,000lbs of fuel and was carried in two tank groups within the fuselage, one forward and the other toward the rear. The forward group contained tank numbers 1 and 2, plus a small collector box which carried 244lbs of ungauged fuel. Tank number 1 carried 9,060lbs of fuel while tank number 2 held 8,182lbs. The rear group of tanks were numbers 3 and 4, and were fairly evenly matched in their capacities. Number 3 tank held 7,931lbs of fuel and tank number 4 held slightly less at 7,270 lbs. There was also another ungauged collector box which could hold 229lbs of fuel. The main wing held an internal fuel load of 11,792lbs and could also carry ejectable slipper tanks capable of holding 3,600lbs each. In-flight refuelling was a necessary requirement, and although would not be fitted on the pre-production models, it was

## TSR-2 PHOENIX OR FOLLY?

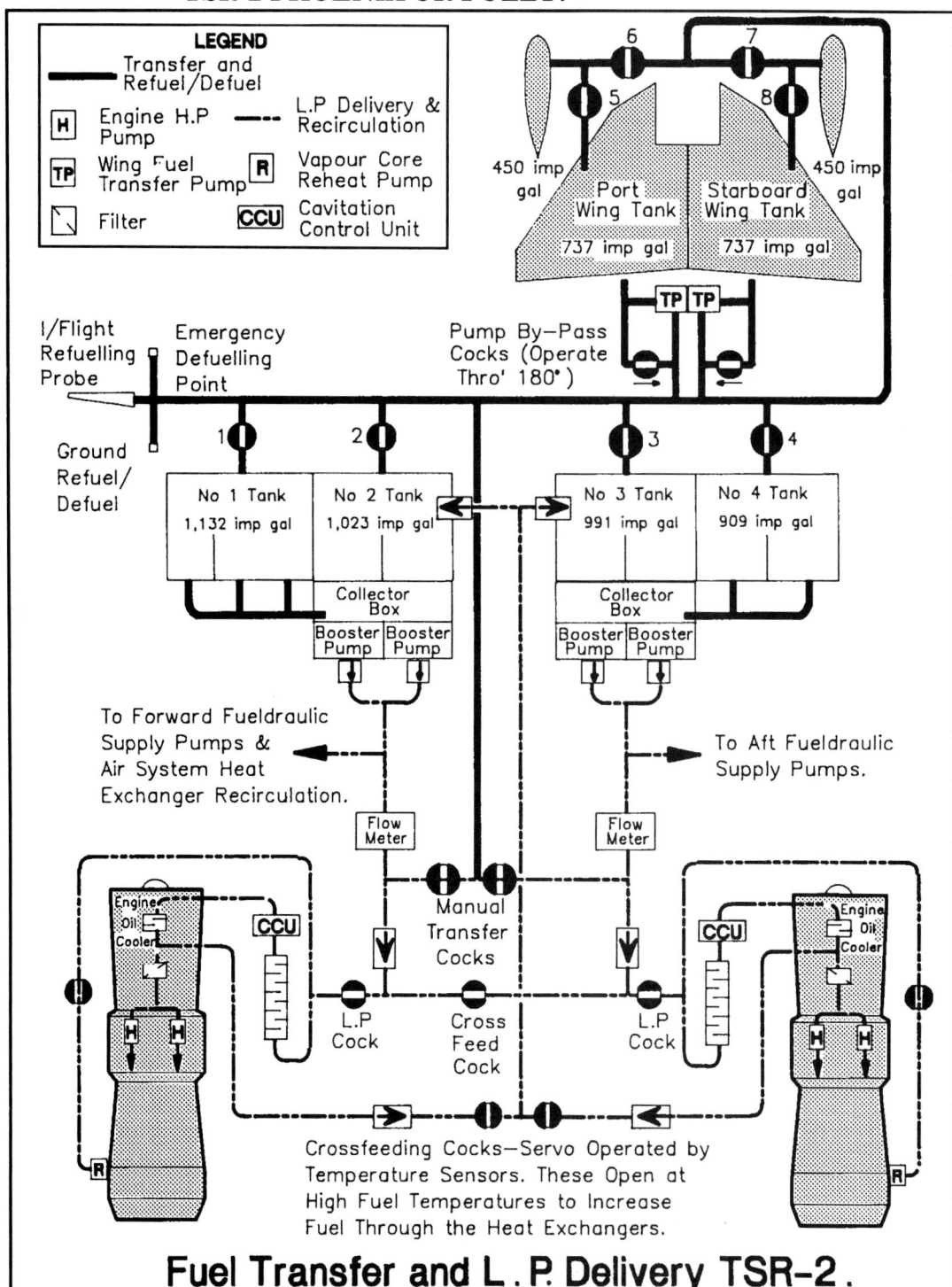

**LEGEND**
- ▬ Transfer and Refuel/Defuel
- [H] Engine H.P Pump
- --- L.P Delivery & Recirculation
- [TP] Wing Fuel Transfer Pump
- [R] Vapour Core Reheat Pump
- [◣] Filter
- [CCU] Cavitation Control Unit

Port Wing Tank — 450 imp gal / 737 imp gal
Starboard Wing Tank — 450 imp gal / 737 imp gal

I/Flight Refuelling Probe
Emergency Defuelling Point
Pump By-Pass Cocks (Operate Thro' 180°)

Ground Refuel/Defuel

No 1 Tank 1,132 imp gal
No 2 Tank 1,023 imp gal
No 3 Tank 991 imp gal
No 4 Tank 909 imp gal

Collector Box — Booster Pump | Booster Pump
Collector Box — Booster Pump | Booster Pump

To Forward Fueldraulic Supply Pumps & Air System Heat Exchanger Recirculation.

To Aft Fueldraulic Supply Pumps.

Flow Meter | Flow Meter
Manual Transfer Cocks
L.P Cock | Cross Feed Cock | L.P Cock

Crossfeeding Cocks—Servo Operated by Temperature Sensors. These Open at High Fuel Temperatures to Increase Fuel Through the Heat Exchangers.

### Fuel Transfer and L.P. Delivery TSR-2.

It was essential that the fuel levels in the tank groups within the fuselage were kept equal to ensure the centre of gravity was maintained. On early flights this operation was carried out by the navigator/observer but on later aircraft it would be managed by the automatic flight control system. *(Reproduced by Author)*

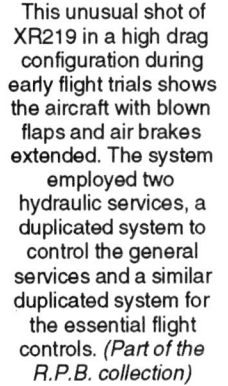

# TSR-2: THE PLANE

intended that aircraft going into service would have an air-to-air refuelling system with a retractable probe situated on the port side of the fuselage just below the cockpit.

**Fuel Transfer.**
With nearly 20 tons of fuel being carried in the long fuselage of the TSR-2 it was important some means of automatic fuel control was included in the system, to ensure the centre of gravity was maintained within very fine limits. The teams at Preston were responsible for making sure this system worked properly, and had no less than four attempts at getting it right before they succeeded. Both the pilot and navigator cockpits had indicators for showing not only the status of the aircraft's fuel supply, both in the fuselage and wing tanks, but also the all important centre of gravity indicator. As explained earlier, fuel played an important part in maintaining the aircraft's centre of gravity and this was controlled by an Auto-Fuel-Balancing system, or AFB. AFB depended greatly on fuel gauging and flow-metering and to ensure the integrity of the C of G, these systems had to cross check within fine limits.

There were certain conditions where manual intervention was necessary to maintain the delicate balance. This mainly came about during essential fuel transfer and the flowmeters had been bypassed. This was particularly so during in-flight refuelling and required the navigator to oversee the operation.

During take-off, with full fuel loads, the C of G emphasis was to the rear, and during this period it was essential the AFB be switched off and the crossfeeding disengaged to ensure fuel was drawn evenly from each group. Once airborne the

This unusual shot of XR219 in a high drag configuration during early flight trials shows the aircraft with blown flaps and air brakes extended. The system employed two hydraulic services, a duplicated system to control the general services and a similar duplicated system for the essential flight controls. *(Part of the R.P.B. collection)*

# TSR-2 PHOENIX OR FOLLY?

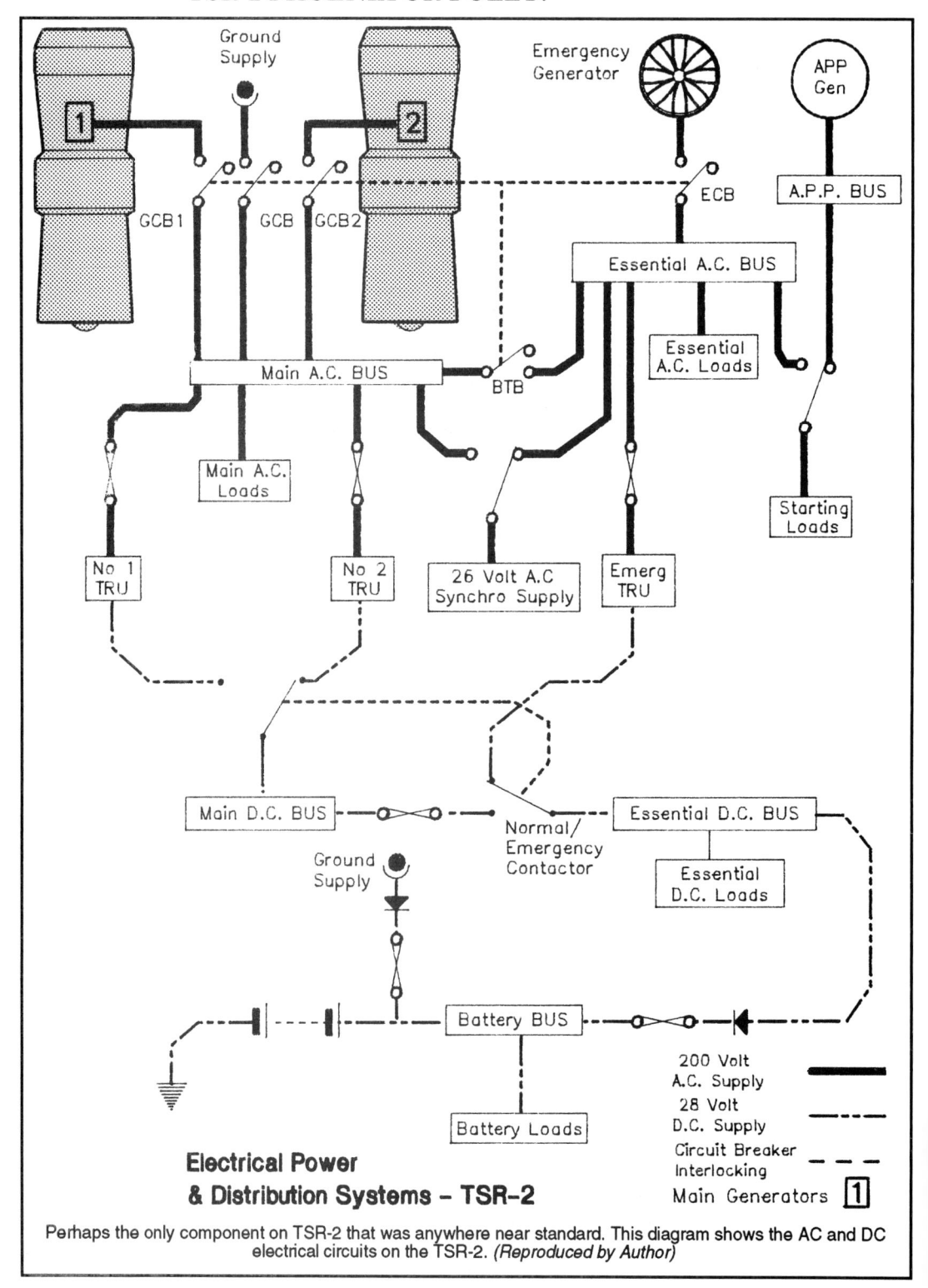

**Electrical Power & Distribution Systems – TSR-2**

Perhaps the only component on TSR-2 that was anywhere near standard. This diagram shows the AC and DC electrical circuits on the TSR-2. *(Reproduced by Author)*

# TSR-2: THE PLANE

AFB system was switched on, and providing there was sufficient space in the forward tanks, fuel was drawn from the rear group to the front group, which moved the C of G forward. When sufficient fuel had been transferred from the rear group, fuel was then taken from the wing tanks, which in turn were topped up by the drop tanks, if these were being carried. When wing and drop tanks had transferred all their fuel, the AFB ensured that both forward and rear group contents were kept virtually equal but with a slight emphasis on the front group. This balance was maintained until the aircraft was in the landing phase, when the emphasis would then be moved to the rear tanks to ensure the C of G was aft.

On the early aircraft, XR219 and XR220, the Auto-Fuel-Balancing system was not fitted and so the cross checking and monitoring had to be done manually. The navigator was mainly responsible for this and accordingly was equipped with more relevent instruments than the pilot. Two additional, but vital indicators for the navigator were 'Fuel Remaining' and 'Flight Time Remaining' and it was essential these cross checked. Today in the modern military aircraft, manually maintaining the C of G appears to be a thing of the past. Fly-by-Wire techniques ensures aircraft stability relative to the C of G.

**Electrical Systems.**

The electrical system used in TSR-2 was perhaps the only internal system that appeared to be anywhere near a standard component. Each of the Olympus engines had an A.C generator, one normally being in service whilst the second was kept in standby-mode. The power generated was 200v, 400c/s, 3-phase, A.C. This was quite sufficient to supply all the electrical needs of the aircraft.

There were a large number of components that required a 28v D.C supply, and these included the fuel auto-balancing, head-up-display computer, the majority of the radio equipment and also the forward looking radar, plus a number of others. This supply was taken from the A.C supply and converted to DC through Transformer Rectifier Units, TRU's. It was designed so that one TRU was kept online with a second in standby-mode. In the event of a total failure in both generators and TRU's, there were emergency components built into the system which would come on line when both, or either failed.

The Ram-air cooled generators were driven by an accessory drive gearbox with a constant speed drive and starter unit. The output from each generator varied from 30kva in the hot ram air stream, to 55kva in the cool stream. The emergency generator drive was quite different and was by a constant speed hydraulic motor, fed from either of the hydraulic system services. There was an Auxilary Power Plant, (APP), generator, but this was only used to supply power for engine starting and had a relatively low output rating.

Should there be failure in the main A.C system all switching to the emergency mode was done automatically. The pilot was informed of this by an an indication on the main console. Although this particular function was automatic in its operation, the pilot did have the opportunity to select the Emergency mode himself simply by selecting an emergency switch. In the case where he felt it desirable to switch back to the main system, he could do so by resetting the Emergency Generator switch.

The D.C supply used similar methods of distribution to the A.C system, but the emergency supply, in the event of both TRU's failing, was from a 22 volt battery with a 7amp-hour capacity. Should a failure occur it was essential that the battery loads were kept to a minimum to conserve energy in order that the aircraft could return to base.

# TSR-2 PHOENIX OR FOLLY?

*XR219 landing after its first flight in September 1964. The picture shows the large brake parachute designed to give TSR-2 a short field landing capability. There was concern that the size of the chute would impose stresses beyond that for which the aircraft structure was designed. (BAe)*

**Stopping TSR-2.**
TSR-2 employed the use of a Brake parachute in the landing phase to help reduce the landing roll-out, with either an automatic or manual system of deployment. The chute was housed at the extreme rear of the aircraft between the jet-pipes with the chute door forming part of the rear fairing. The normal mode of operation was automatic and was initiated by the pilot pulling a handle placed on the port instrument panel, usually during the approach phase of the landing. This single pull deployed a 6ft primary chute. The main chute would then be deployed by undercarriage switches on touchdown. The chute was 28ft in diameter but could be reefed or reduced to a diameter of 16ft, by a peripheral cord. Should the de-reefing facility be required one of two automatic cutters could be used to sever the cord, but only at speeds below 140 knots. If attempted over this speed, a pilot switch would operate, breaking the electrical control circuit. Once the speed was bled off and with the aircraft coming to taxying speed or a standstill, the main chute could be jettisoned normally. TSR-2 was the first aircraft to use a reefing effect brake chute. This was designed for use in heavy crosswind conditions of up to the permitted limit of 35 knots, although drifting would be expected, especially in wet runway conditions. The chute, when used in the de-reefed mode, would permit use in crosswinds up to 20 knots.

For short field landings it was recommended the automatic chute deployment was overridden by pulling the chute release handle twice. This was to ensure the time delay between the chute door opening and the main chute deploying was kept to an absolute minimum. It would also ensure deployment should the automatic sequence fail. A recommendation was also made that the chute was de-reefed for maximum braking effect. Once the primary chute had been deployed, it could not be jettisoned, so during an overshoot condition the pilot would select full power and simply burn the chute off. For his next landing there would be no brake chute to assist with the landing rollout and therefore the aircraft would require a longer braking area.

**Developments in America.**
At this point, it is probably worth looking at what projects aircraft manufacturers were involved in in the United States in addition to the F-111 project. Probably the biggest and most expensive of these projects was the North American XB70 Valkyrie. Manufacturing techniques used on this aircraft were also being pushed to

# TSR-2: THE PLANE

*The North American XB-70 Valkyrie. Conceived around the same period as the TSR-2, the XB-70 flew only a few days before XR219. Designed to replace the B-52's but was dropped because of the huge running costs. The aircraft was later used in the cancelled American SST programme. (North American)*

the limits as this was to be a Mach-3 bomber to replace the now ageing Boeing B52s. The XB70 was a large delta wing aircraft of some 105ft wingspan. The outer third of the wings could be folded down to 25 degress for low altitude supersonic flight or up to 65 degrees for high altitude Mach 3 flight. The construction of the XB70 had to take into consideration the need to travel its entire mission, a distance of some 7,600 miles, without refuelling and with varying types of stores, at its design speed of Mach 3. This meant new metals and different manufacturing processes to combat the high temperatures expected at such high speeds.

The wings of the Valkyrie were very thin and covered with brazed stainless-steel honeycombed sandwich panels, which were welded together. The leading edges of the wings with the same honeycomb sandwich, were attached directly to the front spar. The fuselage was of semi-monocoque construction, built mainly of titanium from the nose to just forward of the wing and stainless-steel over the wing. Power was from six General Electric YJ93-GE-3 turbojets producing some 31,000lbs thrust with reheat. The first XB70 made its maiden flight on September 21st 1964, just three days before TSR-2. In all the XB70s made some 95 flights, logging over 185 hrs. The project was set back when an F-104 chase plane was in collision with the second of the Valkyrie aircraft. The remaining aircraft continued to do a number of test flights mainly to investigate flight parameters for the proposed American SST. The project was cancelled mainly due to the huge costs of the programme, with the single surviving XB70 finally being assigned to a museum.

Although I have mentioned the XB70, this aircraft never was nor would have been considered as a TSR-2 replacement, for the simple reason that the XB70 was a high altitude bomber and not capable of carrying out the roles specified in the British Operational Requirement. Also its cost would have been far in excess of any TSR-2. However it is interesting to see the sort of development that was going on in the USA at the same time as we were building TSR-2.

### Considering Maintenance.

Very few people who worked on the aircraft and its systems, believed the designers had considered the implications of men in the field having to work on the aircraft especially in inclement weather conditions. The Bristol Siddeley engineers who were at Boscombe Down during the early engine runs prior to the first flight, felt too

## TSR-2 PHOENIX OR FOLLY?

much time was wasted when, even trying to cure simple problems, they would be totally frustrated by badly designed access and inspection panels. A ground crew would spend many hours changing parts that should have taken only a few minutes to replace. With the introduction of the Tornado, some of these problems appear to have been overcome and access to the engine is far easier. However engineers on Tornado are still quick to point out that access to other areas appear to be equally as bad, and the avionics bay still requires a person of small stature to work in the area. Design is essentially a compromise between widely varying requirements. To keep TSR-2 and Tornado together at 700 - 800 knots IAS and provide access panels for maintenance required a pretty good compromise, which many felt had been adequately achieved.

**The Schedule.**
An exacting programme had been drawn up for the TSR-2 which included a detailed list of the various tasks each development aircraft was to carry out. Although progress of the whole development programme would be dictated by the availability of development airframes, it did give those who would be working on the test programme some indication of the progress expected over the forthcoming months. In the event of a serious problem arising, then the whole programme would become affected by delay, especially during the early period when only a single aircraft was available, which in turn would tend to magnify any time lost. However, with the number of aircraft that were scheduled to join the programme it was anticipated that any time lost by such delay would be clawed back relatively quickly. The first flight of a TSR-2 had been anticipated as early as March 1963, and had this date been met, then this book may never have been written, but delayed as it was, until September 1964, it was still expected that by early May 1966 all nine development aircraft would be ready for flight testing. This then would have given the development team a considerable amount of experience and confidence in the

This study of XR219 during its first flight, shows the powerful full span blown flaps in action. *(BAe)*

## TSR-2: THE PLANE

project, and also by this time the pilots from the Aeroplane and Armament Experimental Establishment would have started their assessments.

XR219, being the first aircraft, was of course to do the initial trials of the aircraft's basic systems. These should have included aircraft stability and control, the operation of the aircraft's services which took in the undercarriage, flaps and airbrakes, appraisal of engine handling, position error measurement, auto-stabilisation and fin control, a limited amount of measurement of structural and component temperatures, limited checks on the electrical and other systems, and finally the intake performance, including buzz and cone schedule optimisation. The whole programme was not only set back by the engine problems, but also by the extended delays in getting the undercarriage to retract properly. These delayed XR219 being ferried to Warton where major modifications could have been carried out quickly.

The actual equipment deficiencies in XR219 were considerable, but had been anticipated. Back in February 1964, when the first flight test programme had been drawn up, a number of problems had been highlighted. The engines that were to be used in the taxying trials and first flight were designated 320X, indicating that the performance of these engines was lower than that of eventual production units. This was the effect of a less efficient turbine. Also the early marks of engine did not have either water injection equipment or the anti-icing facility, due to these pieces of equipment not being flight cleared in time. The engine intake cone was to be fixed at 17° cone angle during early flights and would be altered manually as the flight test programme progressed. The testing of the automatic cone operation was scheduled for the sixth aircraft and it was therefore not essential it be fitted straight away on XR219. There was further concern with the Graviner Fire Detection System, in that the fire wire control box required a modification before the aircraft would be allowed to fly to a ram air temperature of 110°. As it was not anticipated this temperature would be exceeded during the first flight, it was therefore agreed XR219 would be allowed to fly without this modification. The standard fire bottles initially fitted, required a modification that would allow the ram temperature limit to exceed the imposed limit of 115°.

The all important Automatic Flight Control System, AFCS, was to be omitted in XR219 and although the majority of deficiencies were to be rectified at the aircraft's first 'lay-up', it was doubtful whether the AFCS itself would be fitted in this aircraft at all. This was due to a number of reasons, but mainly attributed to problems with the friction damping in the tailerons and fin jacks. It had been found that there were high frequency instabilities in the control circuits and although there

The taileron piston jack. The tailerons together provided pitch or, independantly to provide roll. The piston itself was duplicated in case of a single failure. This particular unit is still used by the RAF to teach students in engineering at the RAF's No1 School of Training, Halton.
*(Author)*

# TSR-2 PHOENIX OR FOLLY?

was a form of friction damping in the units themselves, because it was causing an unacceptable phase lag, a different form of damping was required. The spigot bush on the fin jack also required redesigning before the AFCS could be used on subsequent aircraft. With the flying control surfaces being at the tail end a linkage was required between the tailerons and fin in order to coordinate the turns, this would give some fin input during a pilot induced turn. The test pilot would be able to alter the amount of input by use of mechanical gearing which he could adjust to give

## *The Main Components of TSR-2*

1. Pitot tube.
2. Terrain Following Radar.
3. Head-Up Display.
4. Lower UHF Aerial.
5. Pilot's Martin-Baker Ejection Seat.
6. Navigator's Ejector Seat.
7. In-Flight Refuelling Probe.
8. Sideways Looking Radar.
9. Water Seperator.
10. Oblique Camera.
11. Liquid Oxygen Tank.
12. Avionics Bay.
13. Auxiliary Pitot Tube.
14. Noseleg.
15. Forward Fuel Tank Group.
16. Doppler Aerial.
17. Inertial Platform.
18. Variable Intake Cone.
19. Airborne Auxiliary Power Unit.
20. Engine Intake Relief Doors.
21. Upper Anti-Collision Light.
22. HF Notch Aerial.
23. Starboard Wing Fuel Tank.
24. Olympus 320 Engines.
25. Water Injection Tank.
26. Accessory Gearbox.
27. Hydraulic Reservoirs.
28. Hydraulic Accumulator.
29. Port Wing Fuel Tank.
30. Main Undercarriage.
31. ILS Aerial

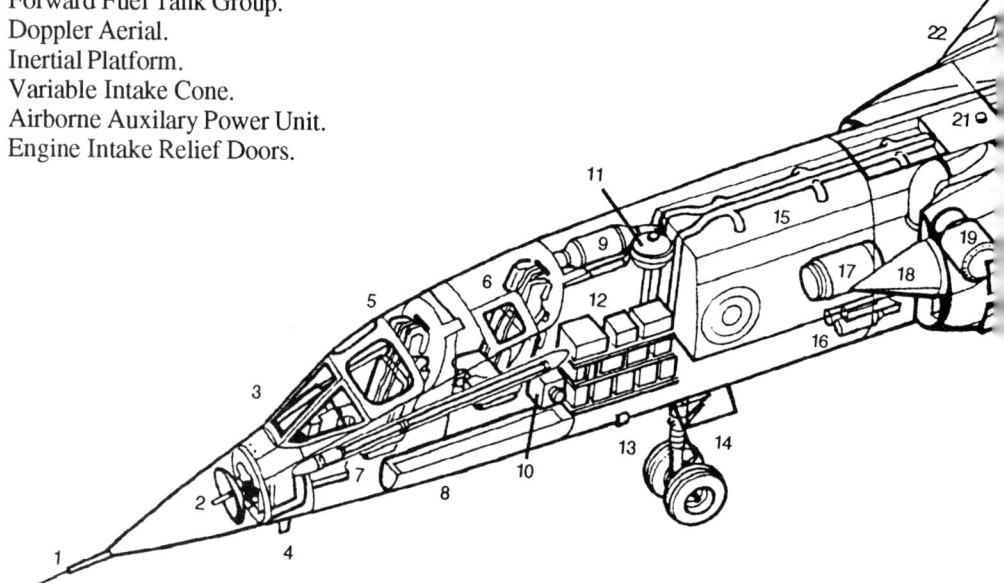

## TSR-2: THE PLANE

him optimum handling. In later aircraft the connection would be fixed and any changes in the input would be handled by the Automatic Flight Control System which would apply the necessary changes appropriate to the flight conditions. XR219 was also to be without auto-stabilisation and auto-stiffening and so would not carry out any initial assesment of this facility.

The complicated fuel system with its method of crossfeeding and the important centre-of-gravity controlling system had also not been fully completed. Should any

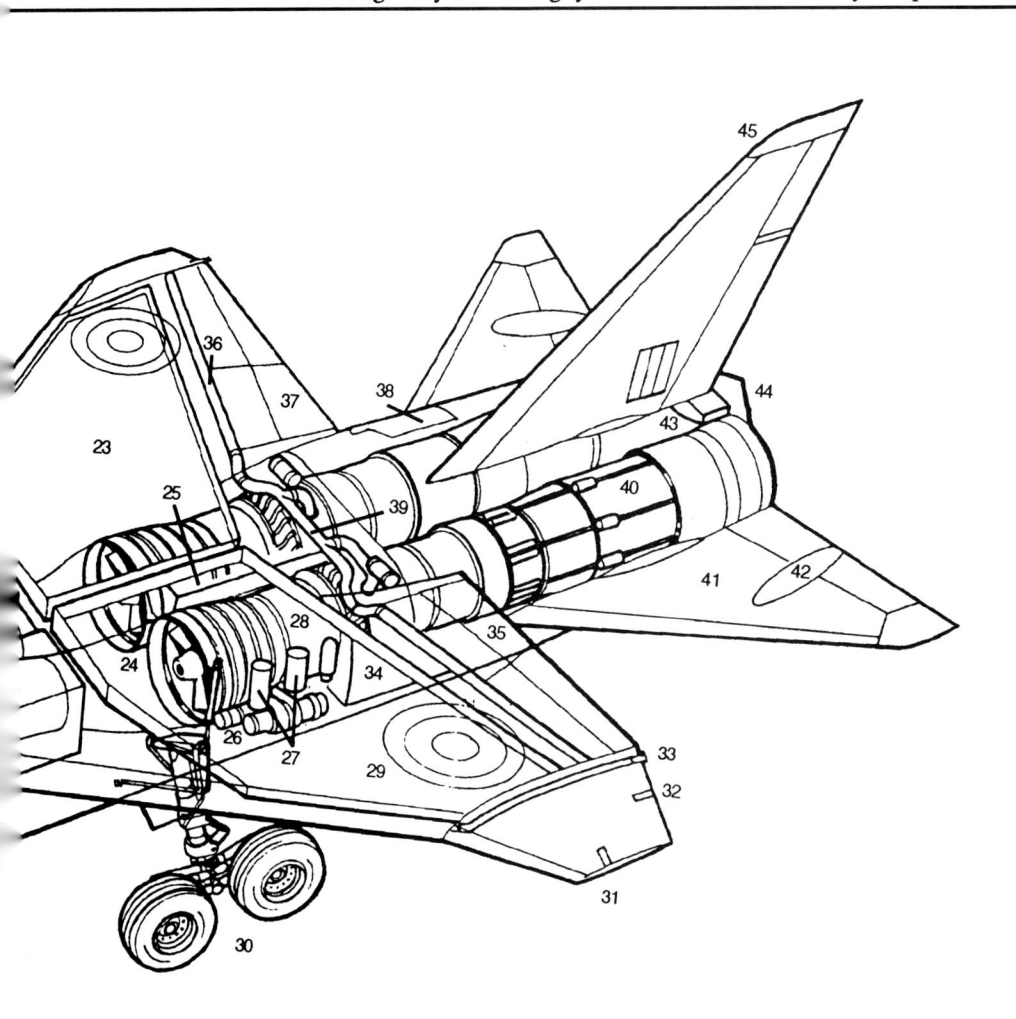

32 ECM Aerial.
33 Wing Tank Fuel Vent.
34 Rear Fuel Tank Group.
35 Port Flap.
36 Flap Blowing Duct.
37 Starboard Flap
38 Upper Air Brake (Starboard)
39 Flap Blowing Cross-over Pipes
40 Olympus Jet Pipes.
41 Taileron.
42 Taileron Trim Jack.
43 Brake Chute.
44 Brake Chute Fairing.
45 UHF - VHF Aerial.

balancing be required on the first few development aircraft, then it would have to be carried out manually by the navigator, who would also be closely monitoring the fuel situation. Particular attention would have to be made with the crossfeeding especially during simulated single engine failures or an actual engine failure, when the crossfeeding cocks would have to be left open in order that the 'live' engine could take fuel from both tank groups. Because there was a problem with the fluid level float switches in the wing tanks, for the first few flights these tanks would not be usable. This in itself would not affect the first flight as the weight restriction put on the aircraft was 70,000lbs and meant they would not be needed anyway. It was imperative, however, that they be modified as soon as possible so as not to slow down the progressive build-up of flight testing with full fuel loads.

Early problems had been experienced with the design of the brake parachute and modifications had already been instigated. The original chute had a manual system whereby the pilot would need to pull the chute release handle twice, once in order to deploy the primary drogue and a second pull to deploy the main chute. On XR219 this whole operation was to be carried out by a single pull on the chute handle, but would be modified to the twin pull action for later aircraft. By normal aircraft standards the chute on TSR-2 was huge and tests had shown that its size was perhaps greater than necessary. It was therefore feared the braking effects of the chute would impose a considerable strain on the airframe beyond that for which it had been designed. To overcome this on XR219 the main chute diameter would be restricted by use of a permanent reefing chord, this would be in addition to the cuttable one already fitted.

The air control valve on the Constant Speed Drive, CSD, had been manufactured in aluminium bronze which imposed an operating temperature limit of 450°. Although its effect on early flights would be little, it was necessary the material be changed to steel in order to allow greater operating temperatures, therefore enabling subsequent aircraft to exceed the set limits below 25,000ft of Mach 1.3 at 20,000ft and Mach 1.1 at sea level. To ensure the actual limits were not exceeded in XR219, the navigator could monitor a CSD air valve temperature gauge in his cockpit.

The navigation equipment fitted in XR219 was to be nowhere near as sophisticated as that described in the chapter covering that subject in respect of definitive TSR-2s. In fact XR219 was to use a simple TACAN unit as found in most military aircraft. Radio communication was to be limited to a UHF, although XR219 would also have a standby UHF set fitted. The VHF and HF radios were scheduled to be fitted in XR221 onwards.

Most of the test measurement and instrumentation was to be housed in the weapons bay which was to pose a number of problems, particularly when the aircraft would be operating at high ram-air temperatures. The equipment would require a cold air supply which would normally have been supplied by a Cold Air Unit for the Reconnaissance Pack. Unfortunately such a unit was not yet available and therefore cold air would have to be vented from the equipment bay cooling supply. The equipment bay itself was to pose a number of problems, mainly in that shelving in the bay would be subject to vibration levels far in excess of those originally expected under conditions of extreme turbulence and likely to become dislodged. It was therefore imperative a redesign of this area should be instigated straight away to ensure complete safety. The weapons bay bomb doors would have to be locked and disconnected from its switch in order to ensure no test equipment was accidently ejected.

# TSR-2: THE PLANE

There had already been problems with the airbrakes and therefore during the initial flights the brakes were to be restricted to 50° of opening. The designed angle of 70° would be acheived as further flight development was undertaken. If one studies photographs of XR219, particularly in flight, it may be noticed the air brake doors appear to be slightly ajar. Because of problems with the locking device the doors were left unlocked and appeared to be slightly open.

For all the deficiences in XR219, the aircraft was pushed far beyond the original parameters laid down for initial prototype flying. This rapid progress was based on the fact that the initial handling inspired great confidence and as the aircraft exceeded all expectations, the opportunity was taken to explore its capabilities further.

These then are the facts as they happened. However, a great deal of work had already gone into preparing a flight test programme for each individual aircraft and a detailed study of the programme itself will give some indication as to what lay ahead for the TSR-2, before its projected in-service date sometime in 1968. One of the most important phases during the development of any aircraft is achieving the all important Certificate of Airworthiness release. This particular milestone for the TSR-2 had been earmarked for the end of August 1967. Up until this time each airframe would be allocated a special certificate which would allow it to carry out specific roles purely as a development aircraft. This assessment was largely based on the Project Pilot's judgement of the aircraft.

**XR220 K0.2.**
XR220 had been the next aircraft due to fly and had it done so without any major problems on the scheduled day in April 1965, then the aircraft would have stayed at Boscombe Down for the next seven or so weeks. During this time XR220 would have undergone initial handling and flight trials before being flown back to Warton. It was then expected that from around mid May XR220 would have been involved in flight envelope clearance trials which would have studied flutter, stability, engine performance, and such measurements dealing with structural temperatures. The early part of 1966, XR220 was scheduled to undergo vibration measurement testing before being laid up between May and June. Quite possibly by then Bristols would have had the fully modified engines ready for fitting in XR220 and no doubt the undercarriage would have required some attention prior to the next set of tests. This was to involve doing load measurement trials on the undercarriage itself. A further lay-up was then scheduled before XR220 was to carry out low speed handling trials to check buffeting and stalling characteristics. By March 1967 the A&AEE trials people at Boscombe would require the aircraft to make further assessments, handing it back to BAC in order that the necessary modification could be made on XR220, before the next set of trials work was carried out.

XR220 was also to be the aircraft to carry out the trials work carrying external stores and if one studies the aircraft today at the Aerospace Museum at Cosford, blisters on the engine intake covers can be seen just to the rear of the auxilary intakes. These blisters were to hold the cameras to film the work that was to be carried out, particularly when the stores were ejected. This work had actually been scheduled for the beginning of May 1967 and was to carry on right through until approximately mid July 1968. XR220 would have been involved in assessing the aircraft's flying capabilities throughout the flight envelope whilst carrying various stores.

## XR221 K0.3.

The third aircraft in the programme was XR221 and had been allocated to carry the avionics equipment. By March 1965 the airframe had already been fitted with the majority of this equipment and ground integration trials were underway at Weybridge. Don Bowen had run a successful ground sortie on the equipment in the aircraft and had found it to be performing beyond all expectations with very few problems expected with the equipment once the aircraft was in the air.

The initial trials were to involve the navigation and attack systems and by the end of 1965 it had been hoped to hand the aircraft over to a Boscombe Down crew in order that they could carry out an assesment of the equipment by flying a number of sorties, using the nav/attack systems in particular. By the end of January 1966 the equipment was scheduled for further development and at the same time work on the terrain following radar was to be undertaken. After this, the aircraft was to be handed back to Boscombe at around September 1966 for further assessments. From October 1966 onwards, XR221 was to have been involved in further development of the terrain-following system, CWAS development and the all important Reversionary Mode trials, before once again, being returned to the A&AEE people for further assesment. Then in June 1967, XR221 would have been back in the hands of BAC ready to undergo auto-navigation and auto-attack systems development. Having completed some 280 hours of development flying, XR221 would have continued well into 1968 with the integration of the auto-navigation, auto-attack and terrain following systems.

By the end of June 1965 it was expected that the majority of problems likely to affect control and stability in TSR-2, would have been sorted and any modifications required would start to be passed down the line for inclusion in the next generation of aircraft.

## XR222 K0.4.

Therefore when XR222 was to have been ready for its first flight in July 1965, a number of these modifications would have been included. However, as XR222 was to explore the longitudinal handling and auto-stabilisation effects, this just might have uncovered some problems. By April 1966 it was hoped this aircraft would be able to move on to the next stage of the development work already planned, which was height, heading and speed locks, mode integration in the subsonic envelope and then by November 1966, that sufficient work would have been carried out so that the Boscombe pilots could once again use XR222 for an assesment on these aspects of the aircraft's development. Providing Boscombe found no new problems, XR222 was then scheduled to continue with Auto-ILS trials including extension to full envelope. This would include a further month with A&AEE pilots between July and August in 1967, before being laid up for further modification work and the fitting of extra equipment. By the begining of October, XR222 was to start development for the manual terrain following and in March 1968 would begin the auto-terrain following trials. Moving on to November 1968, it was hoped that XR222 would be ready to begin auto-pilot trials carrying external stores.

## XR223 K0.5.

The schedule for XR223 appears to only go as far as the begining of June 1967. The first flight for this aircraft was destined for September 1965 and was to include preliminary bombing trials and trials with both canopies removed. After a short lay-up in February 1966, XR223 was to begin further trials that would include checking

# TSR-2: THE PLANE

the air systems and structure temperatures which would last until approximately June 1966, when it was anticipated XR223 would be laid-up for equipment fitting in readiness for the tropical trials that were to begin around December 1966. These trials were to be undertaken by the pilots from the A&AEE and were scheduled to last until June or July of 1967.

### XR224 K0.6.
XR224 was due to make its first flight in October 1965 and had been scheduled for Airframe and Engine Performance Trials. This work was to continue through until December 1966. During this time certain periods had been set aside for assessment trials by the A&AEE pilots at Boscombe Down. By February 1967, after a two month lay-up for modifications, the aircraft was to be made available for any further necessary development tasks. In September 1967, XR224 was to undertake further trials involving the carriage of external stores, through until mid February 1968.

### XR225 K0.7.
The electrical, fuel and hydraulic system trials were scheduled to begin when XR225 became available, which it was anticipated would be in January 1966. These trials were to continue right through until the end of July when XR225 was to be assessed by the A&AEE. The 'in-flight' refuelling was to be an important step, vital in that it would be the testing ground for the 'Fueldraulic' system and any effects this method of refuelling may have on the all important Centre-of-Gravity monitoring system. The refuelling trials were expected to have taken place between September 1966 and mid January 1967. Following a successful trials period XR225 was then to be laid up for major modifications and the fitting of the Reconnaissance Pack. The pack, described later on, was to be assessed on the aircraft starting at the beginning of March and was to go right through until the first few weeks in June 1967. After a further lay-up for modifications, further trials were to take place, again using the Reconnaissance Pack and by March 1968 it was hoped to begin flying sorties using the pack to its fullest.

### XR226 K0.8.
The penultimate development airframe was to be XR226 and this aircraft was scheduled to begin its trials work toward the end of Feburary 1966. Although its job was particularly important, it was even more so in that this was to be the airframe responsible for proving the navigation and attack systems and for carrying out the radio trials. Between February and August 1966, the aircraft was to do the initial nav/attack integration trials and air alignment. Providing this important stage passed without problem, the aircraft was then to carry out navigation systems development in the subsonic role and at the same time undergo development for the attack manoeuvre. Following the completion of a successful subsonic trial, the next phase was to take the aircraft to a high altitude and perform in the supersonic role. It was hoped this work would be completed by the March of 1967, when XR226 was due to be laid up for modifications after which, it would be required to complete a number of important proving sorties.

### XR227 K0.9.
The final aircraft in the development batch was of course XR227. So far no mention has been made to weapons carrying and relevant trials work. However,

this was to be the aircraft that was to conduct the development of the weapons packs, both internal and external. XR227 was scheduled to join the development flying team toward the end of April 1966. The initial development testing was due to last until the end of December 1967 and would have concentrated on stores which were to be carried internally, of weights from 950lbs up to 2100lbs. The work was to have included the carriage of the stores in various flight envelopes and their release under differing conditions. There would have been the usual release of the aircraft to A&AEE pilots during this period for their individual assessments. By January 1968, having completed several successful missions with internal stores, the development phase was to move on using the aircraft's hardpoints on the wings to assess the carriage and release of bombs and the firing of various rocket packs. It was hoped this work would have been completed by the end of 1968.

As can be seen from this programme it was anticipated that all the development aircraft would have been available for flying duties within a very short period. As Roland Beamont had already pointed out, once the initial snags had been overcome on XR219, the testing and development of the aircraft speeded up very quickly, and there appeared little doubt that this pattern would have continued. This is not to say the development programme would have been trouble free and experienced no setbacks to cause further slippage in the schedule, but from those initial flights no major problems were expected There was little doubt that by 1968, the Royal Air Force would have been receiving its first aircraft at the proposed Operational Conversion Unit at RAF Coningsby in Lincolnshire.

## CHAPTER THREE
## A Major Step Forward.

The reliance upon electronic equipment and similar aids for flying and navigating aircraft has perhaps been the most significant trend since the second world war, with one of the biggest breakthroughs being the ability to fly aircraft in conditions of extreme bad weather with pinpoint accuracy. It has also given the pilots of our military aircraft the ability to deliver their weapon loads with that same level of accuracy, always a most desirable element in this particular facet of aerial warfare. The equipment being developed for TSR-2 represented one of the biggest advances in avionics the aviation industry had known.

The varied amount of avionic equipment the industry was developing for TSR-2 was both considerable and unequalled in its day. No other military aircraft within the United Kingdom had been designed with such revolutionary equipment as on-board digital computers and similar devices that would enable it to fly missions with very little pilot intervention and at altitudes which would avoid detection by enemy radar and all of this in extremes of bad weather. Although at the time a number of electronics firms were working on individual pieces of equipment, such as Blue Shadow, Green Satin and later Blue Silk, (Blue Silk being a refined Green Satin). No aircraft so far had been specifically designed to carry such equipment, or was to be equipped with an assortment of this equipment, that would ensure the accuracy demanded in OR343. Aircraft such as the Lightning, those of the V-Bomber force, and not forgetting the Canberra, were often grounded because of the lack of such equipment, whereas in TSR-2 virtually every aspect of the main flying and navigational functions had some electronic input.

The Operational Requirement from which the TSR-2 was designed and built insisted the aircraft have the ability to fly deep into enemy territory at virtually tree top height and deliver its weapon load with pin-point accuracy during a single pass. For a pilot to be able to carry out such a manoeuvre without any assistance in the form of electronic aids was virtually impossible. Equally the navigator was unable to carry out his role because of the inability to pick up reference points under these same conditions. It was therefore only logical that the TSR-2 employ such advanced methods of flight control and navigation that would enable the aircraft to carry out these tasks with as little visual reference to its immediate surroundings as possible. To do this would not only draw upon the assistance of a number of specialised firms whose expertise in radar and electronics was well known, but perhaps more significantly it would impose a huge financial burden on the project. Eventually this would culminate in an aircraft that would be a considerable number of years ahead of its nearest rival.

It must be understood that what follows is only a very brief outline covering some of the major components in the TSR-2 in order that the reader can form some idea not only as to its complexity, but also as to the intended role of the aircraft.

**Electronic Warfare.**
There would be no single item of equipment that would carry out the demanding role of navigation, and no one role would be performed by a single piece of equipment. The whole weapons platform would therefore be designed using various pieces of equipment, which in turn would continue to function even when other equipment within the system became unservicable. The system would

comprise of a central computer, terrain following radar, autopilot, doppler, inertial platform and a sideways looking radar. Coupled to these would be a number of peripheral systems that would navigate the aircraft extremely accurately without the use of any external navigation aids. Its accuracy was such that after the aircraft had been flying for 1 hour, any error, on average, would not exceed more than 1 1/3n.m and considerably less on the final stage of the attack. With many of the inputs also being used for other equipment, it was estimated that should there be a single failure of a basic sensor, its affect would only result in a minimal effect on the aircraft's performance and with only a slight degradation in service. Such an accurate navigation system would ensure a weapons delivery of such accuracy that previously would have been considered impossible.

It was also important to consider that as TSR-2 was using an inertial system it could not be jammed by outside interference as the aircraft would be a completely self contained unit. As well as specialised companies being involved, the Royal Aircraft Establishment and the Royal Signals and Radar Establishment would also be very much part of the development team. Some of those companies involved included Decca, Elliott Automation, Ferranti and Rank Cintel. Elliott Automation, for example, had been manufacturing avionics for some time and their contribution to the development of new equipment for military aircraft in particular, was already well known.

**Computing in Aircraft.**
At the time Elliott's were working on a navigation system which employed the use of a digital computer. The work was in direct competiton with the GEC Research Laboratory at Stanmore, which was receiving government funding for a project known as DEXAN. It was found however, that although DEXAN was fairly well advanced, it fell considerably short of the requirement demanded by the BAC Electronic's Division. The computer that was being proposed by Elliotts for the TSR-2 was the VERDAN, which in fact was already in production, acting as a guidance system in an American ballistic missile. The computer itself had originated from Autonetics, a division of the North American Aviation

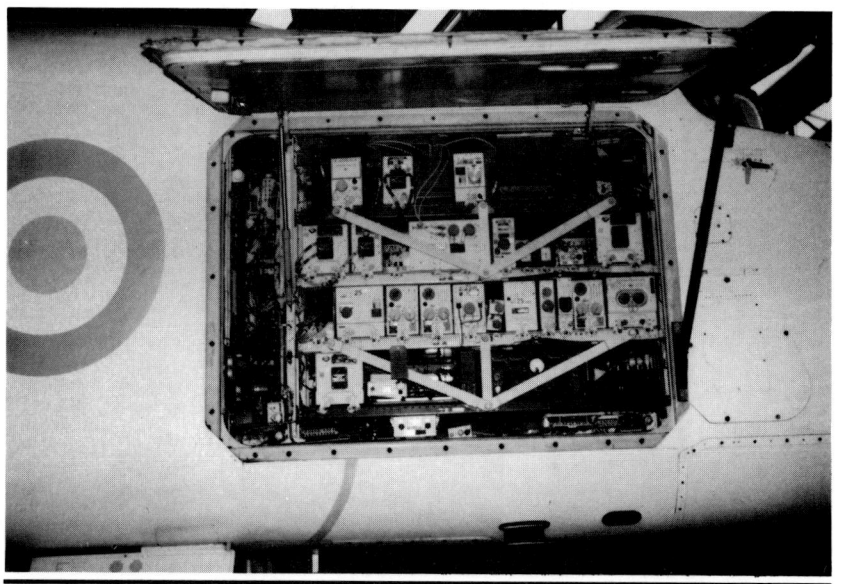

The large avionics bay on TSR-2 was to the rear of the navigators cockpit, taking up both sides of the fuselage. The bay was to carry most of the avionic equipment that would have made TSR-2 unique. This picture shows the bay on XR220 rebuilt by the Cosford Aerospace Museum to give some idea of what the aircraft would have looked like.
*(I. Frimston)*

# A MAJOR STEP FORWARD

Corporation, and Elliott's were manufacturing the unit under licence in Britain. In order to carry out the more demanding role of flying and navigating an aircraft under such extreme conditions, this computer had to be extensively modified. The computer software used a digital form and in comparison with today's standard, had quite a small memory of only around 2k, which was stored on hard disk. The computer itself used a 'digital differential analyser', which used digital representations for the analogue quantities. The unit would receive input from various peripheral pieces of equipment which was then cross checked with a predetermined course which had previously been fed into the computer, and any deviation would be corrected. It also had an attack mode, which is covered in more detail later on.

**Accurate Navigation.**
The aircraft's navigation and attack systems were very closely integrated and relied upon data which had been pre-fed into the digital computer. There were two Verdan computer systems, using information from the Decca doppler and the Ferranti inertial navigation systems, that made up the attack mode. A feature of their flexibility was that they would allow the navigator to select a 'revisionary' mode, which would permit continued operation but with reduced accuracy, should any of the other components fail. The digital computers themselves did not communicate with each other, nor could they be made to do so, but were both kept online performing the same functions. As the specification had insisted on a certain amount of redundancy it was therefore felt the system would require a back-up in case the primary unit became unservicable. The aircraft's mission plan was pre-plotted on paper-tape. This tape would contain all the relevant information for the particular mission involved and included all the necessary check points from where, if required, the navigator could update the navigation system and take out any errors. The tape was fed into both computers, so each unit was capable of controlling the mission in the case of unservicability in the other.

**Doppler.**
The doppler was to be supplied by Decca. This in itself would supply the system with accurate ground speed and drift information, which was essential for pin-point navigation. Both the navigator and the Verdan Central Computer System, (CCS) would receive information from the doppler. Some output could also be used to drive the Moving Map Display and this equipment is covered in more detail later on. The CCS was required to be operating normally in case of a fault in the doppler or, during those periods when it was outside its normal operating limits, for example, when the aircraft was making violent manoeuvres. During these periods the CCS would use the last values in memory given by the doppler. The doppler required intricate and exacting methods to gather its information, so the system employed two methods of collection.

Doppler operates by sending out a signal in a forward or rearward direction, angled downwards from the aircraft. Taken that some parts of this signal would be transmitted back, the frequency returned would be different to that sent out. This is known as Doppler Frequency. The ground speed is calculated by measuring the amount by which the frequency has increased or decreased, toward or from the point of reflection. A more accurate form of measurement could be made if simultaneous transmissions were made forward and rearward from the aircraft.

To determine the aircraft drift angle the system would use two pairs of beams.

## The principles of Doppler Drift Angle

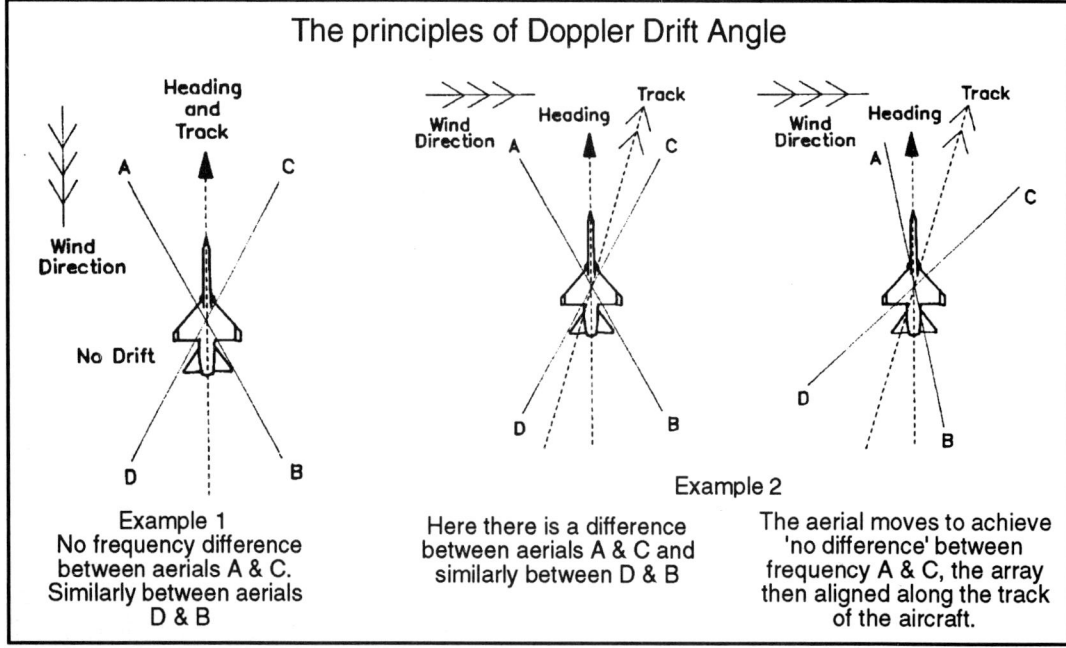

**Example 1**
No frequency difference between aerials A & C. Similarly between aerials D & B

**Example 2**
Here there is a difference between aerials A & C and similarly between D & B

The aerial moves to achieve 'no difference' between frequency A & C, the array then aligned along the track of the aircraft.

One pair would be transmitted forward and downward on the port side and rearward and downward on the starboard side. In conjunction the second pair would transmit in the opposing directions on the opposite sides. The beams would be sequenced so that once every second, pulses along one pair of beams would alternate with pulses along the second pair. The drift was determined by having the aerials able to move through the required angle to establish identical returns from left and right aerials. With no drift, the return from the left and right aerials would be the same. With drift being experienced, there would be a difference in the doppler frequency measured by the two forward aerials, also by the two rearward. This difference actuates the pairs to move to establish identical returns from left and right aerials. This angular movement is measured against the fore and aft axis of the aircraft, the heading. Thus the track of the aircraft would be established and with groundspeed converted into distance by time, the position determined. This method did require a very stabilised aerial array and therefore used output from the inertial platform.

The more complex principles of the doppler system to obtain an accurate drift angle. Another method was to transmit a beam downwards from the aircraft. The Tornado transmits three downward beams to obtain the same accurate information.
*(Author with assistance from Brian McCann)*

### Inertial Platform.

Because the production inertial platform would not be ready in time the first five pre-production aircraft were to be fitted with a type 100. This lacked the capability expected of the production type 200 models which would form the basis of a dead-reckoning inertial navigation system. Platform stability was essential, and was provided by three single axis gyros, one defining azimuth and two defining the vertical in the N/S and E/W planes respectively. Accelerometers were then placed to measure acceleration vertically and horizontally in N/S and E/W directions. A simple gyro system as described would maintain the platform at a constant orientation in free space, whereas what is required is a platform maintained to the local vertical on a rotating earth. In order to achieve this the Verdan computer was programmed to compute the distance travelled round the earth's surface taking into account the airframe's acceleration, the coriolis accelerations and the earth's rate of

# A MAJOR STEP FORWARD

The RAE at Farnborough became involved in the development of various pieces of equipment. This rear cockpit mockup was used mainly for equipment handling and ergonomic design. *(RAE)*

rotation and the latitude of the aircraft. The results of these calculations were then converted to voltages which were applied to torque motors on the gyros so as to cause them to precess at the necessary rates.

The platform and gyros were required to operate in a strictly controlled, heated environment. Therefore the equipment was housed within a container where such control could be easily maintained. This was achieved by using a thermostatically controlled system of individual heaters for the platform and its associated gyros. Before each flight the navigator had to go over a complicated set of procedures to ensure all the navigation equipment was initiated correctly and that it was functioning within its designed parameters. The platform itself, having been pre-heated, was normally aligned in either the 'Gyro' or 'Magnetic' mode. There was a provision so that the platform could be aligned 'in-flight', but as this was time consuming during flight, the normal practice would be for it to be done with the aircraft stationary on the ground.

**Initialising the System.**
The navigator would switch on the power which would activate the essential heating systems and at the same time give a rough alignment. This particular sequence was time consuming and could take up to 5 minutes, but depended upon the present temperature of the gyros. Once the indicator lamps were lit informing the navigator that the gyros had reached their operating temperature, the alignment switch would be set to ALIGN-1. At this stage the navigator would set the local vertical and the 'Y' axis, which was aligned with the navigator's compass. Again, when he received the appropriate indication that his had been done he would select ALIGN-2 and carry out a similar procedure using the 'Gyro' and 'Magnetic' compass in order to reduce to a minimum any errors. This whole procedure could

# TSR-2 PHOENIX OR FOLLY?

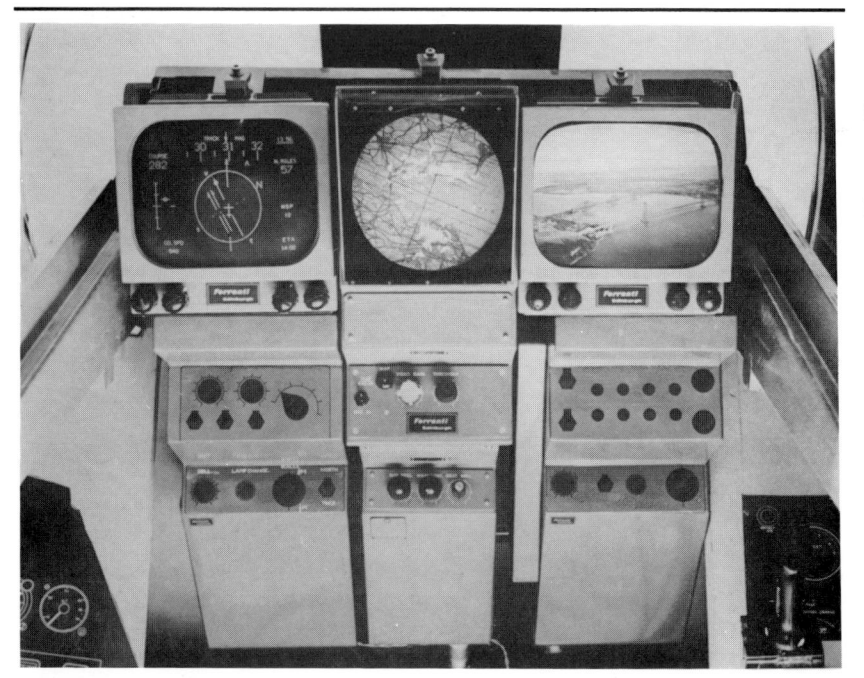

*Ferranti were very much involved in the avionics equipment for the TSR-2. This picture depicts Ferranti's concept of what the rear cockpit would have eventually looked like. Such a display would not look out of place in a modern jet bomber.*
*(GEC Ferranti)*

take anything up to 25 minutes, however, the longer it took in excess of 20 minutes with the aircraft stationary, the more the level of accuracy within the system increased. If such a procedure did have to be carried out whilst airborne, to obtain the same level of accuracy, the aircraft had to fly straight and level for at least 20 minutes with the CCS and doppler fully operational.

**Looking at American System's.**
The Royal Signals and Radar Establishment, (RSRE), at Malvern, was to become very much involved in the specification of the radar requirements for the TSR-2, as laid down in the Operational Requirement. The technical specifications for the various radar systems were to be written by the Establishment and then sent out to a number of companies in order that they could submit tenders for the work.

One can perhaps appreciate the complexity of such systems and the RSRE were similarly concerned at the amount of development work needed for new equipment in the TSR-2. During the initial stages RSRE members visited Texas Instruments in the United States, as this company was also involved in similar projects and at the time was working on a forward looking radar that was eventually to be used in the F-111. The company was hoping to secure a contract from the RSRE to supply a radar system for the TSR-2. RSRE's assessment of such systems concluded that whilst the American equipment did have certain advantages, it was insufficiently advanced to have saved Britain any considerable time or money. Therefore the RSRE decided it was best to continue with the work being done in this country.

**Terrain Following Radar.**
In 1955, Cornell University had been awarded an American Defence contract to study ground hugging radar. However this contract lapsed without the Americans using the information and was taken up by Vickers and subsequently Ferranti. From this came the Ferranti Monopulse System. At the time it was unique and only one

# A MAJOR STEP FORWARD

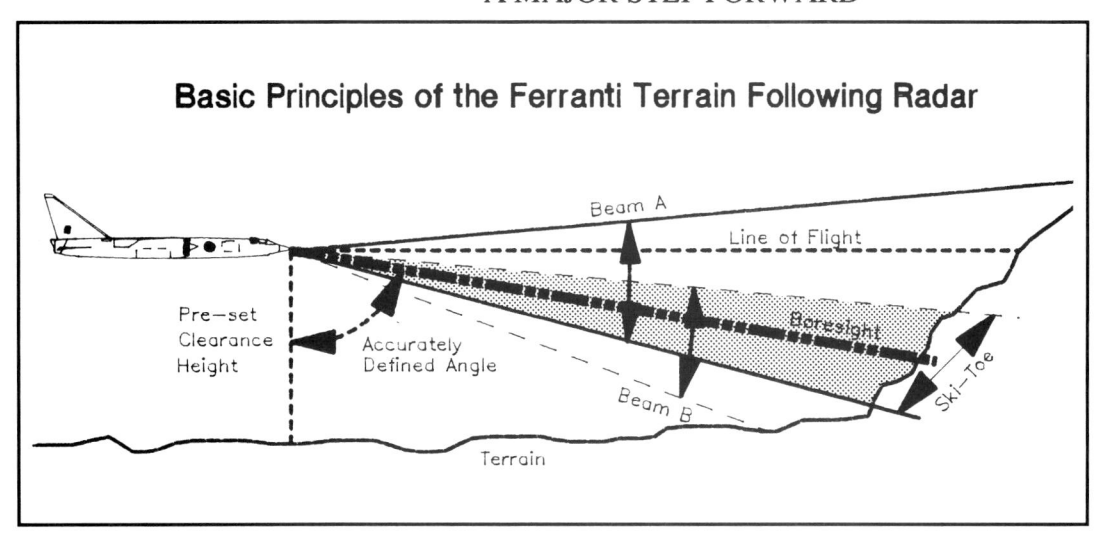

**Basic Principles of the Ferranti Terrain Following Radar**

*The basic principles of Terrain Following Radar. (Author with assistance from Brian McCann)*

other company was known to be developing a terrain following radar in America. Together with the on-board computer, Doppler, sideways looking radar and autopilot made this a formidable weapons system.

The contract for the Terrain Following Radar, or TFR, was eventually awarded to Ferranti Ltd. The original requirement had been for a 'Multi Mode' radar system encompassing several facilities. These included the Terrain Following itself, a Ground Mapping facility, Air to Surface Ranging and a Beacon Homing Mode.

The main role of the radar was the Terrain Following Mode which would provide a safe path for the aircraft at minimum height over all types of terrain and water. The range of the radar itself was reputed to be in excess of 40 miles. Basically, the radar scanned the path ahead of the aircraft and sent signals to the autopilot, with output also being presented to the pilot via the Head-up-Display. This was particularly helpful in instances where the aircraft was under manual control. There was also a special display for the navigator which showed the actual radar return signals against a template which indicated the flightpath. Although initially this was provided as a 'confidence builder', the main reason was to give any indication of electronic jamming. The autopilot was handled by the automatic flight control system, or AFCS, which had outputs linked directly to the hydraulic flight controls. The pilot could select a minimum clearance height, between 200ft and 990ft. Coupled to this was a ride control where the pilot could effectively vary the 'g' limit which the steering data demanded. This meant the crew could smooth out the dips and hollows thereby minimising the effects of fatigue.

Another important requirement was that the radar had to operate safely in turning flight, which meant it had to anticipate the flightpath around corners in order to ensure that adequate account was taken of obstacles which otherwise would have appeared suddenly at close range and too late to take avoiding action. Therefore, this particular system was designed to give full performance in turns of 45° of bank at Mach 0.9.

At the time this was a unique system being developed by Ferranti. It was the first time the company had developed a radar system using transistors for all functions, except the main transmitter tube, which was a magnetron. This increased reliability, and also reduced the power consumption and lowered the ambient temperature.

The radar unit weighed approximately 230lbs and was built as a single compact unit, which was to be housed in the nose of the aircraft behind a glass-fibre reinforced nose cone. It was, like most Ferranti radar systems, of the monopulse type which was extremely accurate and could determine angles to within a fraction of a degree. It had the ability to pick up not only the contours of the earth, but also other objects likely to endanger the aircraft's mission in particular objects such as aerial masts or power lines. Another and important requirement was the ability of the radar to recognise when the aircraft was flying down a slope towards water.

Terrain Following Radar itself receives a very poor return from water, especially inland water areas where the surface is usually very smooth. This tends to deflect the signal with the radar receiving very little return. Therefore in instances when this happened it was essential for the radar to recognise the situation so that the necessary steering instructions could be given early enough to prevent the aircraft flying straight into the water. As a consequence a radio altimeter was fitted with a vertically downward transmission to give height data over water. Such an accurate radar enabled the aircraft to fly as low as 100ft over water or flat lands from 350 knots to mach 1+, with no need to adjust the parameters within the radar itself.

The method by which this demanding operation was carried out involved the radar unit itself, an Air Data Computer and the Autopilot, (the ADC was an anologue computer, unlike the CCS which was digital). The radar gathered its information from the single aerial in the aircraft's nose. This used two overlapping beams from a single source, using two separate feeds. Although at the point where the two beams converged the effective beam width was sharply reduced, it did considerably enhance the range and angle accuracy. The signals were then sent to the Air Data Computer and through the manoeuvre computer, checked the 'most nose up' attitude, before issuing instructions to the autopilot, thus ensuring the aircraft was flown clear of the ground. At the same time the radar would also pass information to the the pilot through the Head-up-Display, or HUD. The ADC also

*The Terrain Following Radar, manufactured by Ferranti weighed only 230lbs. Designed to fly the aircraft safely across unknown terrain, the radar was said to have a forward range of around 40 miles.*
*(GEC Ferranti)*

had input to the autothrottle for the two Olympus engines. This ensured that when demands were made by the Autopilot, the appropriate throttle adjustment would ensure sufficient engine power was available to carry out the manoeuvre.

A radar system such as this was quite revolutionary, for up until now normal radar had been limited to merely presenting a built up picture of the scanned terrain, leaving the crew themselves to manoeuvre their aircraft around any obstacles affecting the flight path.

**Ground Mapping.**
One of the other features of this radar system was the Ground Mapping mode. This provided the navigator with a ground picture of the land ahead. The range depended on the chosen scale and the height of the aircraft, but in general it could map up to ranges of 100 n.m. The navigator could superimpose a topographical map of the area over the radar picture which would enhance the display by giving general information or allow him to take accurate fixes under all conditions.

He would also use the Beacon Mode. This part of the TFR had a range of up to 100 n.m and would interrogate a suitable beacon, with the signals received by the navigator being passed to the pilot as steering instructions. The beacon could either be a ground station as an approach aid, or perhaps in a tanker waiting to refuel the aircraft.

The final mode was the air-to-surface ranging. This had a proven ability to lock-on at maximum ranges in excess of 7 nautical miles, and could even be locked on to runway intersections. This mode would be selected by the pilot at the last fix point before the target, which would normally be when there were about 30 miles to run. The mode flies the aircraft through the delivery manoeuvre to the point of automatic weapon release. It is interesting to note that it was only the air-to-surface ranging that could be used with the terrain following mode, with no other mode having the ability to engage simultaneously.

**Head-up-Display.**
Looking round the cockpits did not immediately reveal this to be an unusual aeroplane, even though there were a number of systems new to the standard military cockpit. These included the Head-up-display, (HUD) and Moving Map display. As previously mentioned the HUD was part of the terrain following system and therefore received its information from the TF radar. The HUD itself was manufactured by Rank Cintel, and presented several types of information to the pilot. These included a Flight Director display, an Attitude Indicator, provision for weapon aiming and limited speed and height information. The complete display was projected directly onto the pilot's windscreen, unlike that of the Buccaneer for example, which was displayed on a secondary reflector plate. By ensuring the display was adjusted according to the pilot's line of sight at infinity, and making the display appear as though superimposed on the windscreen, the pilot was able to maintain a constant lookout. Once the brilliance of the display had been adjusted to suit the pilot's needs, no further manual intervention was required. Any additional adjustment was done automatically to ensure the same brilliance was maintained relative to the changing background.

**Automatic Flight Control.**
The Automatic Flight Control System, (AFCS), supported two separate, but important functions. Firstly it provided autostabilisation and autostiffening and secondly the autopilot system. Autostabilisation was operated through the tail

surfaces and damped out any small oscillations in the aircraft. Aircraft oscillations were detected by three rate gyros the outputs of which, after suitable processing, were used to operate small hydraulic jacks to add a small additional movement to the pilot's controls.

Autostiffening is a slightly different problem. At speeds in excess of 1.5 mach, directional stability in the lateral plane is insufficient, and over 1.7 mach it becomes negative. To correct this, an accelerometer was used to measure the amount of sideslip, and counteracted it with additional input to the autostabilisor, which in turn instigated the necessary fin movement to restore the stiffness in yaw.

The Autopilot System was linked to the Air Data Computer. This used pressure inputs from pitot and static tubes at points in the aircraft's nose. Together with Ram air temperature, these readings could then work out the Mach speed and height. The autopilot system was then capable of holding speed lock, acquiring and holding height, heading and track locks, and also carrying out automatic weapon delivery and performing I.L.S approaches. The speed lock was attained through automatic thrust control, except when the engines were in reheat, or if they were at particularly low settings. It was particularly useful during an ILS approach when the pilot could pre-set a throttle setting and the aircraft would maintain a constant speed irrespective of any manoeuvre. This largely prevented any chance of an accidental stall.

The design had already been set for the rear cockpit on production TSR-2's. A comprehensive, spacious cockpit was something the navigator on the bomber was not used too. *(BAe)*

# A MAJOR STEP FORWARD

The main AFCS controller was situated on the starboard console and contained a number of switches including the 'mode selection switch'. In addition to these were two indicators on the port instrument panel along with a confidence switch. This switch had 24 positions and enabled the pilot to check out the integrity of the AFCS. There were also indicators on the central warning panel immediately in front the pilot. This layout enabled him to control and monitor the AFCS with relative ease.

One of the most important features of the Automatic Flight Control system, was its ability to quickly detect a failure within the Terrain Following system and at the same time, to take immediate action. This system used the forward looking radar, and the radio altimeter and its associated equipment. Input would be taken from two channels and the system selected the one with the most 'nose-up' demand. In the event of total failure, a 'nose-up' attitude was automatically selected, thus ensuring the aircraft was taken clear of the surrounding terrain. However, should there be a single channel failure, a limited amount of response would still be available.

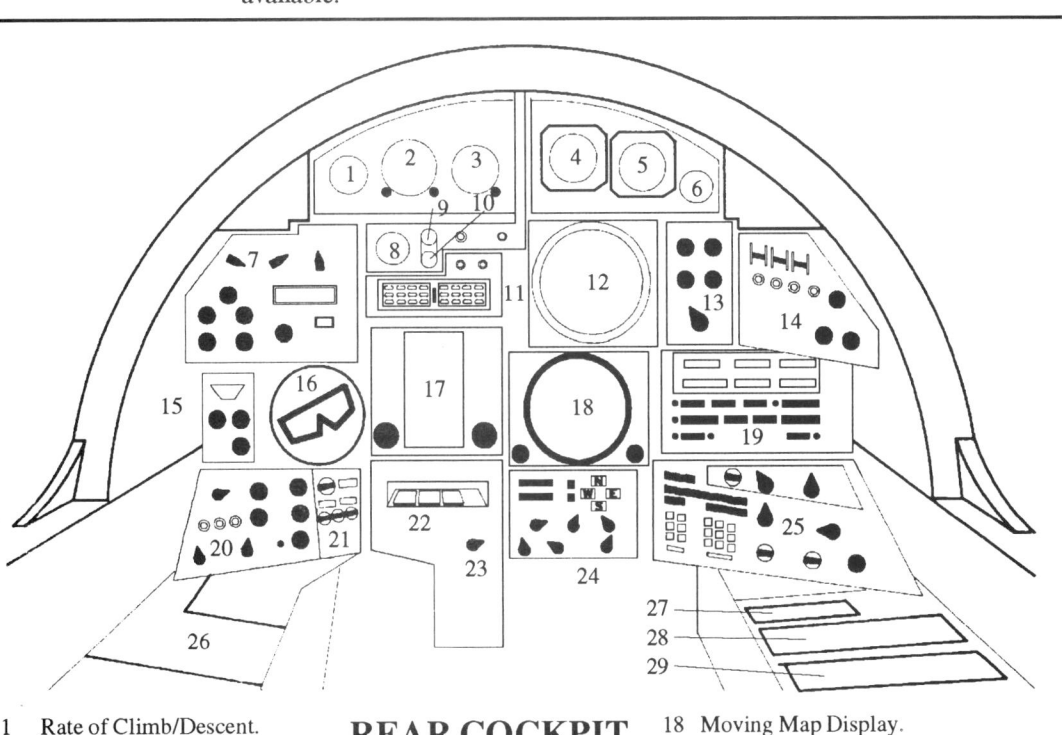

## REAR COCKPIT KEY

1. Rate of Climb/Descent.
2. Altimeter.
3. Air Speed Indicator.
4. Artificial Horizon.
5. Compass.
6. Drift Display.
7. Navigation Radar Monitor and Control Panel.
8. Radio Altimeter.
9. Oxygen Content.
10. Oxygen Pressure.
11. Central Warning Panel.
12. Radar Display.
13. Radar Display Controls.
14. F. L. R. Indicator.
15. Downward Sight Control
16. Downward Sight.
17. S. L. R. Display.
18. Moving Map Display.
19. Mission Display Panel.
20. Recce Radar Control Panel.
21. Recce Camera Selector Panel.
22. Doppler & Computer Advisory Lights.
23. Ground Speed Input Select.
24. Moving Map Control Panel.
25. Mission Control Panel.
26. Linescan Control Unit.
27. Transponder.
28. Navigators Station Box.
29. H. F. Control Box.

# TSR-2 PHOENIX OR FOLLY?

### Moving Map.

The Royal Aircraft Establishments had considerable involvment in developing a great many of the specialised components for the project. One such device was the 'Moving Map' display. This system had originated at the RAE Farnborough, and was later taken over by Ferranti. The idea was for the pilot and navigator to be shown a moving map of the area the aircraft was passing over. The display was simple and easily understood, allowing the crew quickly to pinpoint their position. In order for the map to be sufficiently minituarised and, at the same time retain reasonable resolution, a 35mm colour film was used, onto which were filmed standard aeronautical maps, which were then projected onto a ground-glass screen. As the magnification of the charts had been reduced in order to get a sufficient number on the film, they had to be remagnified when being presented back to the crew. In the centre of the map screen a fixed marker represented the aircraft, and the film was moved across the screen appropriate to the aircraft's path across the ground. The display was driven by the aircraft's central computer which converted the input data into signals which drove the map display. To achieve considerable flexibility in its tactical role, the system had the ability to store a large number of charts on one microfilm. By being able to call on a number of films, there was little restriction in TSR-2's capability to function not only in Europe, but throughout the Middle East and Warsaw Pact countries if so desired.

### Attack Systems.

It was essential that the aircraft should fulfill three distinct attack roles. These being the ability to carry out a nuclear strike in any weather condition, the ability to carry out a reconnaissance mission regardless of weather conditions and to be able to carry out a conventional attack with minimum regard for weather conditions. Placing the nuclear capability first in no way implies that this was the TSR-2's primary role, but is merely used as a starting point to describe its various attack missions.

It was intended that the attack role should be flown as a series of 100 mile legs, and missions were preplanned then fed onto paper tape, with each of the 100 mile legs having a specified start and end 'Fix-point'. The fix-points would be specially selected objects with a good radar signature which could be picked up easily using the sideways looking radar. The tape was then fed into the Verdan computers where the basic navigation, (handled by the doppler/inertial system), was linked to other data sources and the whole mission coordinated by the Digital computer. Using the 'Fix-point' at the end of each 100 mile leg, the nav/attack system could be accurately updated, thus giving the aircraft a correct starting point at the commencement of each leg, with a final update close to the target. All navigational equations and calculations were done by the two on-board Verdan Digital Computers in real time. To ensure TSR-2 fulfilled this role, the aircraft had the ability to change to a reversionary mode in the event of a system failure. This was achieved by using the basic inputs from other computing units within the aircraft to back up the failed system. Although the aircraft could continue its mission, it would be at a degraded level of operation. This would at least ensure not only the aircraft's survival, but also its ability to return to base.

In the tactical nuclear role the aircraft would be flown in the terrain following mode. The route would have been pre-fed into the central computer and would have contained the 100 mile legs along with the target coordinates. At 30 miles to target the weapons release mechanism would automatically be initiated, and the pilot

# A MAJOR STEP FORWARD

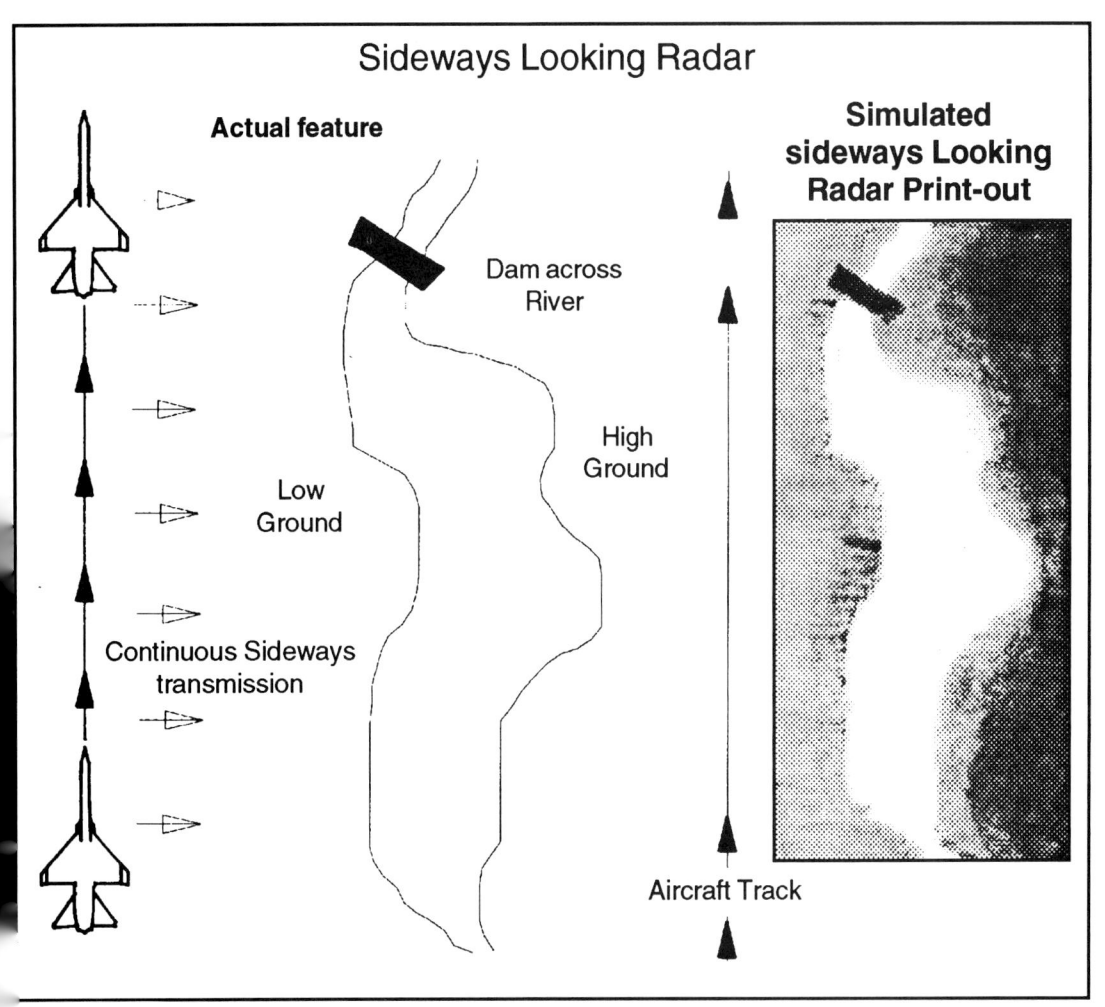

**Sideways Looking Radar**

*Actual feature* — Dam across River, Low Ground, High Ground, Continuous Sideways transmission

*Simulated sideways Looking Radar Print-out* — Aircraft Track

Looking either side of TSR-2, the sideways looking radar had a range of approximately 10 miles. By selecting objects with a good radar return the SLAR would help navigate the TSR-2 extremely accurately and at the same time would form part of the reconnaisance system. *(Author with assistance from Brian McCann)*

would select the Air-to-surface ranging. This would fly the aircraft through the delivery manoeuvre to a predetermined point where the automatic weapons release would carry out its role. This would involve the aircraft being pulled up to a height of approximately 2,000ft, from where the bomb would be released. Such a manoeuvre would accurately 'lob' the bomb onto the target and at the same time allow the aircraft to continue climbing and turning to a safe cruising height away from the nuclear blast and clear of any surface-to-air missile attack.

The strategic role was somewhat different only in that the weapons used were to be the lay-down type. By continuing at tree-top height to the target, a lay-down bomb would be released. This would allow the TSR-2 to reach a safe distance before the bomb exploded. Consideration had to be made in the case of a continuing nuclear attack. It had been proposed to use a number of stand-off missiles such as Blue Water. *(Blue Water was cancelled in 1962 at an estimated cost of £16.5 million.)* When attacking high priority targets deep within enemy territory, TSR-2 would be able to carry various low yield missiles in the weapons bay together with high yield weapons on the wing hard points. The conventional attack sequence took exactly the same format as the nuclear one, using the same navigation and terrain

# TSR-2 PHOENIX OR FOLLY?

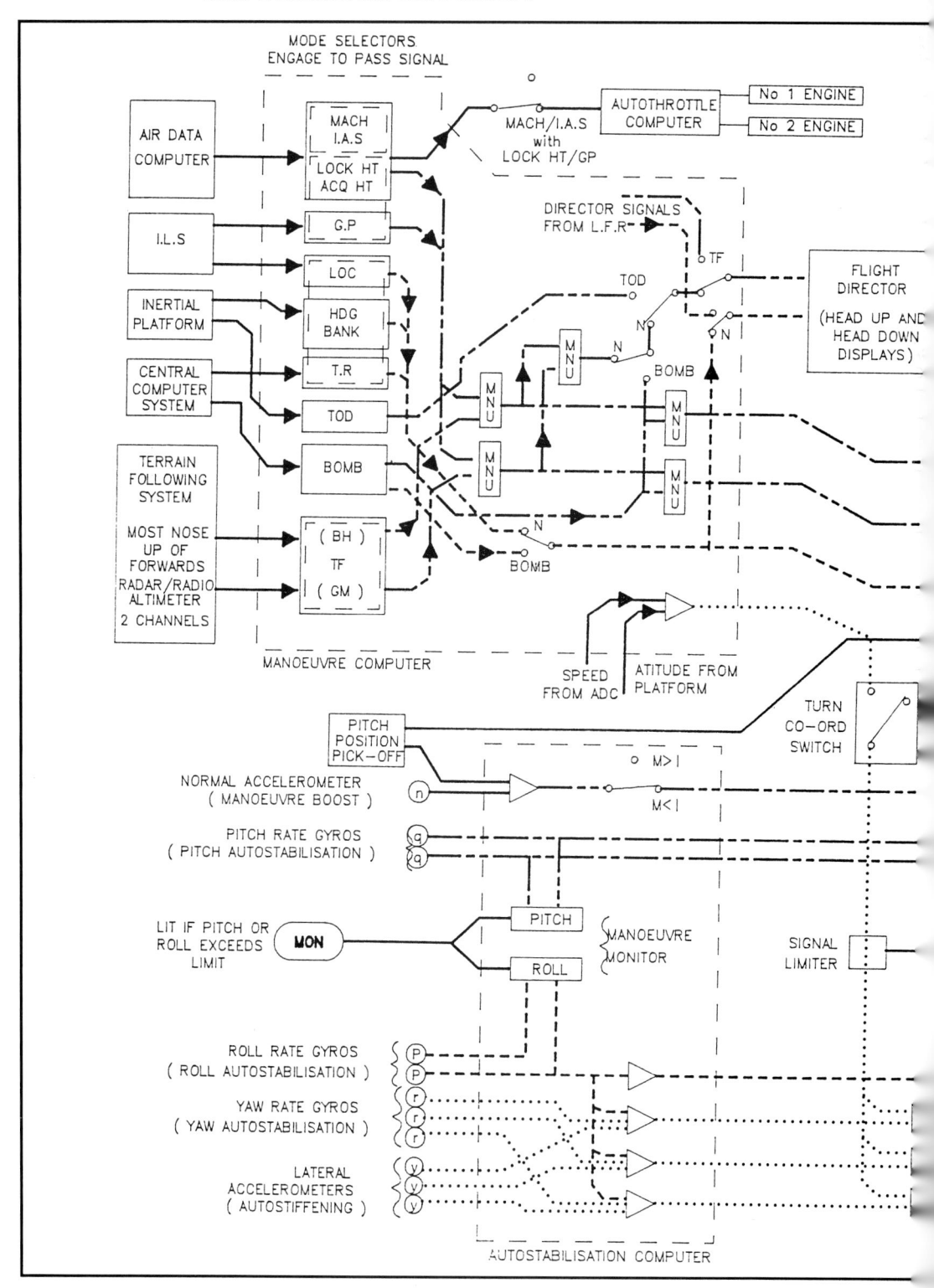

# A MAJOR STEP FORWARD

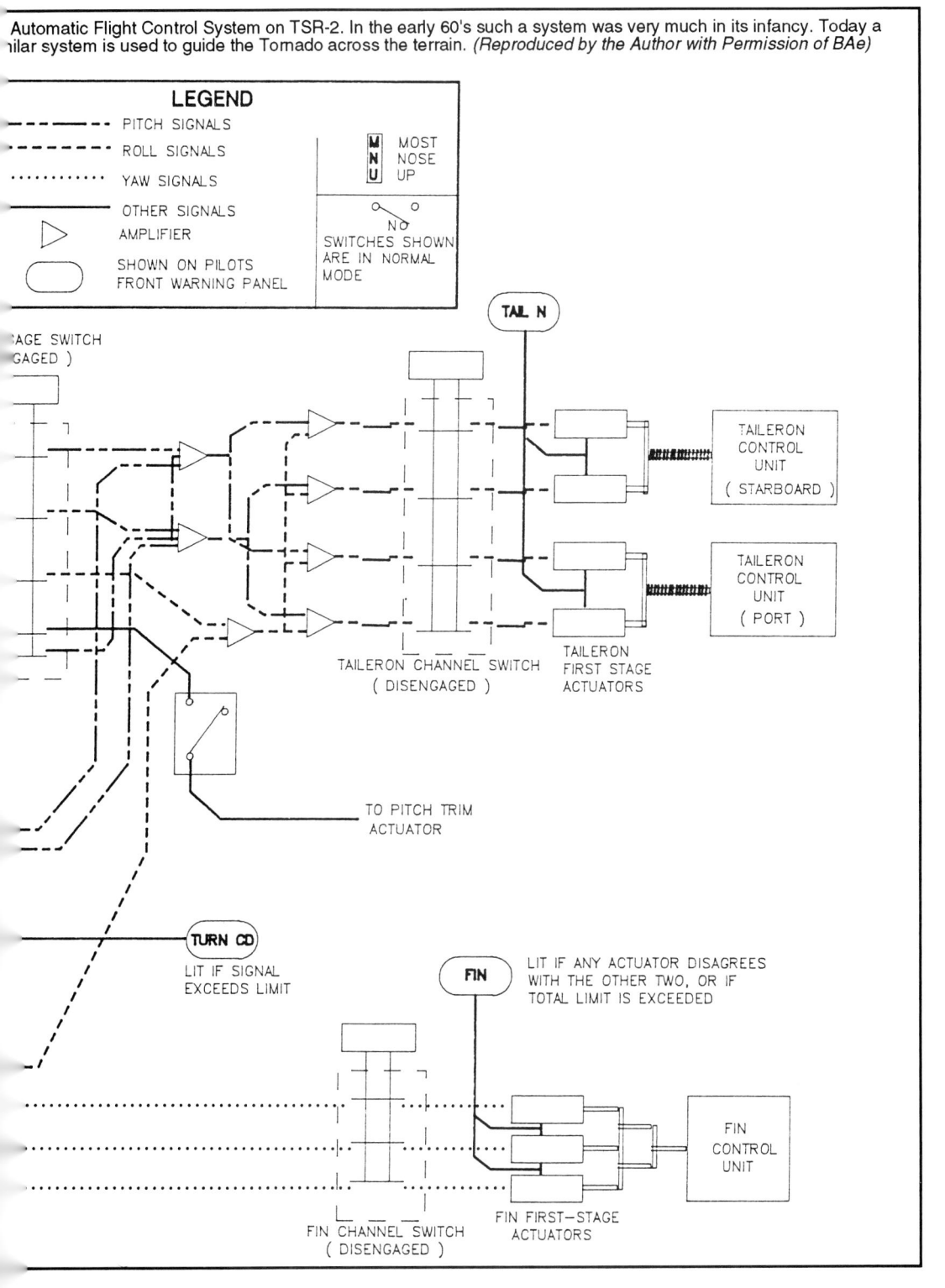

Automatic Flight Control System on TSR-2. In the early 60's such a system was very much in its infancy. Today a similar system is used to guide the Tornado across the terrain. *(Reproduced by the Author with Permission of BAe)*

**LEGEND**

- — - — - — PITCH SIGNALS
- — — — — ROLL SIGNALS
- ············ YAW SIGNALS
- ────── OTHER SIGNALS
- ▷ AMPLIFIER
- ⬭ SHOWN ON PILOTS FRONT WARNING PANEL

M - MOST
N - NOSE
U - UP

SWITCHES SHOWN ARE IN NORMAL MODE

TAIL N

TAILERON CONTROL UNIT (STARBOARD)

TAILERON CONTROL UNIT (PORT)

TAILERON CHANNEL SWITCH (DISENGAGED)

TAILERON FIRST STAGE ACTUATORS

TO PITCH TRIM ACTUATOR

TURN CD — LIT IF SIGNAL EXCEEDS LIMIT

FIN — LIT IF ANY ACTUATOR DISAGREES WITH THE OTHER TWO, OR IF TOTAL LIMIT IS EXCEEDED

FIN CONTROL UNIT

FIN CHANNEL SWITCH (DISENGAGED)

FIN FIRST-STAGE ACTUATORS

following modes. The difference against the planned targets was the method of delivery, in that from the last 'fix-point' the distance to target would be flown visually using the weapons aiming system displayed on the pilot's Head-up-Display. Also in the conventional mode was a method known as 'Targets of Opportunity', which would have allowed the crew to choose a selected target and use the weapons aiming system for weapon delivery.

The navigator's cockpit had a number of large scopes, including one, (as explained earlier), for the moving map display. Another was a television monitor for use with the Matra Martel, (Missile Anti Radar and TELevision), air to surface missile which was to be used as part of the armament on the TSR-2. This missile had a small camera in the nose and pictures were transmitted back to the navigator's scope through a wire. The wire would also transmit signal's from the navigator's joystick whereby he could guide the missile directly onto the target. The missile's range was around 10 miles.

**Reconnaisance.**
One of the main roles mapped out for the TSR-2 was reconnaisance. This particular type of mission would be flown in a pattern very similar to that of a conventional or nuclear attack. The aircraft again would fly its route via the pre-planned 100 mile legs, but this time instead of bombs, the TSR-2 would use its sophisticated reconnaisance equipment at the target area. One such piece of equipment was the Q-band, sideways looking radar, manufactured by EMI Electronics Ltd. The Q-band system, as mentioned before, would be used not only for correcting the doppler/inertial system in its navigational capacity, but also as an electronic aid in the aircraft's primary reconnaisance role. The concept behind this type of equipment was for the radar to look down and sideways from the aircraft. The scanner did not move and therefore could be made long and was mounted in parallel with the aircraft's flight path across the earth. Using a narrow beamwidth was found to give a high resolution picture far better than anything previously seen. The trace, which actually 'painted' the picture on the small display tube, was kept stationary, whilst a long strip of film was moved past the display at a speed proportional to the aircraft's speed over the ground. Irrespective of weather conditions, a complete record of the aircraft's reconnaissance mission was recorded on film, with such detail as bridges, buildings, lakes, rivers and in particular specific contours of the earth. Moving targets left a distinct pattern, similar to that resulting from photographing a moving object with a slow shutter speed. This was a unique system as the aircraft carried the necessary fluids to develop the film in flight. The results could even be transmitted back to base for quick analysis and interpretation. One particular advantage with this type of radar was that due to its method of mounting in the aircraft and its mode of operation, enemy countermeasures could not detect it until the aircraft was overhead.

**Linescan.**
Linescan was another piece of reconnaisance equipment manufactured by EMI Electronics. The Linescan itself was an optical scanning device that swept the ground beneath the aircraft using an electronic beam. The equipment worked in two modes of scan, one being passive where the equipment was scanning in a daylight environment, and the second, an active mode, where a field of light scanned the earth at the same time as the electronic beam. The 'active' mode was usually only selected during night time operation so that a high resolution picture could be

# A MAJOR STEP FORWARD

*XR222 pictured at Duxford after being restored. This aircraft was scheduled to play an important role in testing the avionic systems, especially the Terrain-Following Radar. (IWM)*

obtained. The data could be either stored onboard the aircraft, or transmitted back to a command post using a high frequency data-link. The data-link and its associated ancillary equipment was housed in the reconaissance pack along with a number of cameras.

The individual cameras in the pack each used different focal length lenses and this allowed for reconnaissance over a wide range of altitudes from high altitude to low-level work. Film for each camera was loaded into an individual cassette and contained anything up to 250ft of film. There were also a number of cameras elsewhere in the airframe. One was housed in the nose, while two others looked obliquely from either side of the fuselage slightly below the two cockpits. The reconnaissance pack also contained a video recorder and an optical scanner unit, along with the data-link transmitter. Some of this equipment was temperature controlled and therefore required a cold air unit when the aircraft was operating in high temperatures. This varied type of equipment obviously gave the TSR-2 a great deal of flexibility in the reconaissance role.

The pack itself was carried in the weapons bay and was designed to be easily fitted and removed, in order that all the necessary film and power packs could be serviced separately away from the aircraft.

**Informing the Crew.**

Because of the complexity of TSR-2s equipment, it was essential that both pilot and navigator be warned immediately of any system failure. For this purpose both crew members had centrally mounted warning panels. The pilot's warning panel was directly in front of him just below the line of the windscreen and the navigators situated in the centre of the front panel. Each panel was divided into two classes, red for critical and amber for advisory. The pilot's panel had 48 indicators, 24 red and 24 amber and covered warnings for all the aircraft's systems, except for fire warning displays, which were sited elsewhere in the cockpit. The navigator had 32 indicators, 8 red and 24 amber, and these displayed warnings mainly relating to the nav/attack systems. Each indicator in both panels was illuminated by two bulbs and could be tested by the crew with two press-to-test switches. The displays also had a Day/Night feature allowing the brilliance to be altered. The pilot's panel had two

# TSR-2 PHOENIX OR FOLLY?

'attention getter' lights on either side of the panel, two red and two amber. These were associated with the advisory lamps and flashed when any of the appropriate advisory lamps on the panel illuminated. These attention-getters could be cancelled but would also self-cancel if the problem cleared. Whenever a red attention-getter illuminated it was accompanied by an audible warning.

The flight instrumentation in the TSR-2 was a logical extension of the OR 946 system as used in the Lightning, though with a CSI, Combined Speed Indicator rather than a strip speed indicator. Directly in front of the pilot was the panel containing a compass and attitude display. Below this was the moving map display which was inset into the panel. Not all the dials themselves were easily readable and some of the smaller ones would have required redesigning, in particular those furthest from the pilot's eyes. A modification had already been proposed and was due for inclusion in the pre-production aircraft.

The Head-Down-Display panel contained most of the standard military instruments and included a standard 'roller-blind' attitude indicator, and within this indicator was the Flight Director. A compass was also included, being positioned at the bottom of the central instrument panel: as well as displaying the aircraft's heading, it displayed ILS and drift information. The navigator had a duplicate of this instrument on his panel and was responsible for the synchronisation of both

The front cockpit as it would of looked on production aircraft. *(BAe)*

# A MAJOR STEP FORWARD

compasses. The instrument was able to operate in three modes, and depending which mode had been selected, would gather information from a number of sources.

The ILS information was derived from the appropriate radio beacon and was shown to the pilot by two bars, one a vertical localiser and the second the horizontal glide path bar. The drift information was gathered from the doppler equipment and was shown as a pointer on the outside ring of the compass. If the doppler system became unserviceable, the drift indicator would not be able to function. Mounted on the port instrument panel was a speed indicator with a pointer and Mach scale. The altimeter and Vertical speed indicator were servo instruments, receiving their information from the Air Data Computer. The altimeter had a single pointer and

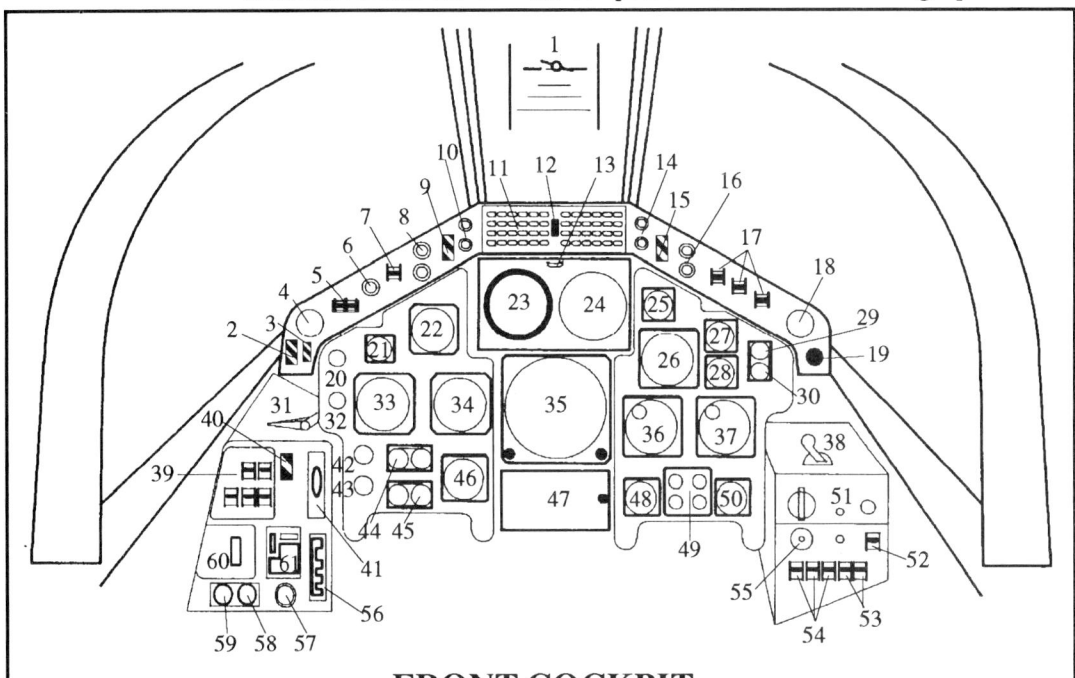

## FRONT COCKPIT

### KEY

1 HUD Display.
2 Nosewheel Extend Switch.
3 Manual Flap Blow Switch.
4 Incidence Meter.
5 Brake Chute Jettison Switch.
6 Brake Chute Door Selector.
7 Brake Chute Reefing Switch.
8 Port Reheat Warning.
9 Port Reheat L. P. Cock.
10 Fire Warnings Port Engine.
11 Central Warning Panel.
12 Day/Night Switch.
13 Turn Indicator.
14 Fire Warnings Starboard Engine.
15 Starboard Reheat L. P. Cock.
16 Starboard Reheat Warning.
17 AFC Selector Switches.
18 Accelerometer.
19 HUD Brilliance.
20 Flap Pressure Indicator.
21 Air Speed Indicator.
22 Combined Speed Indicator.
23 Atitude Display.
24 Nav/ILS Display.
25 Radio Altimeter.
26 Altimeter.
27 Aircraft Skin Temperature.
28 Local Altimeter.
29 Oxygen Content.
30 Oxygen Pressure.
31 Parking Brake Handle.
32 Flap Position Indicator.
33 Standby Horizon.
34 Standby Altimeter.
35 Moving Map Display.
36 Thrust RPM Port Engine.
37 Thrust RPM Starboard Engine.
38 Rudder Pedal Adjuster.
39 Nosewheel Short/Normal Switch.
40 Undercarriage Emergency Switch.
41 Undercarriage Select Lever.
42 Air Brake Pressure Indicator.
43 A. I. L. Gear
44 Brake Pressure Indicators.
45 Hydraulic Accumulator Guages.
46 Rate of Climb Indicator.
47 Radio.
48 Turbine Temp, Port Engine.
49 Intake Cone/Nozzle Position.
50 Turbine Temp, Starboard Engine.
51 ILS Controls.
52 Aerial Selector Switches.
53 Intercom Selector Switches.
54 VHF/UHF Selector Switches.
55 Radio Channel Selector.
56 Flap Position Lever.
57 Flap Indicator.
58 Roll Indicator.
59 Yaw Indicator.
60 Air Brake Standby.
61 Undercarriage Indicator.

## TSR-2 PHOENIX OR FOLLY?

digital readout. One revolution of the pointer represented 1,000ft, with the digital readout being in 50ft increments up to 99,950ft. In the case of the ADC failing, a warning flag would be shown in the instrument and the pilot would have to resort to his Standby instruments. Certain supplementary flight instruments were available, including a Radio altimeter. Although in the early stages of development the radio altimeter, moving map display and doppler system were not fitted into XR219 or XR220, these components were being bench and flight tested in other types of aircraft and did in fact appear in XR223. The two prototype aircraft that were completed, XR219 and 220, did have a self contained TACAN unit fitted.

**Showing the Crew.**
The illumination for the instrumentation in both the front and rear cockpits was another move away from convention. Because of the role TSR-2 was to perform it was important that the instrumentation had a reliable form of lighting. It was felt that with the conventional method of using bulbs, the sudden fusing of a bulb under such conditions was a high possibility. A solution to this problem had been developed in the USA by the Sierracin Corporation. This alternative method of illumination was based on the 'Sierraglo' lamp, as it was known, and whilst providing the same sort of illumination associated with aircraft instruments, did in fact have a number of advantages.

The lamp itself was designed as a complete panel representing the instrument, and inlaid in a tough plastic coating were the appropriate legends of the associated instrument. The legends appeared white in daylight or coloured when in darkness or

TSR-2s Head Up Display was unusual in that it was displayed directly onto the windscreen glass, and may have looked something like this simulation.
*(Author)*

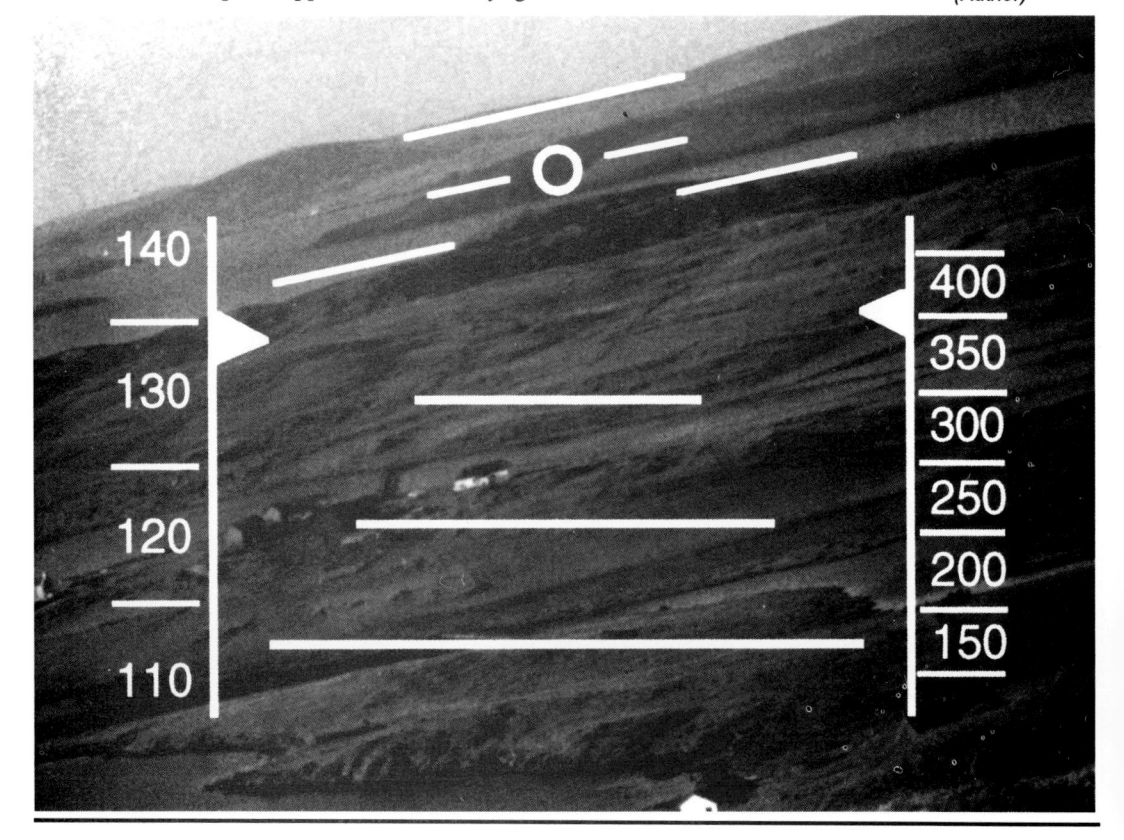

dim light. The panel itself took the form of a thin phosphor sandwich and when an alternating current of electrical power was passed through the phosphor the energy was converted into cold light. When dimmed, it did not vary or lose its effectiveness. As it did not use conventional type bulbs it was not vulnerable to the sudden burn-out problem usually associated with bulbs. Also the need for multiple bulbs and associated wires and holders was eliminated.

**Communication.**
Radio and intercommunication was to be supplied by Plessey (UK) Ltd. The installation comprised of a combined VHF/UHF transmitter receiver with an additional ILS controller for the pilot and a HF controller for the navigator. There was also an emergency UHF radio and intercom system for use in the event of a failure with the main equipment. The equipment had two aerials, one the primary aerial in the fin tip, and a secondary aerial just below the nose of the aircraft, which also served the emergency UHF system. The range of the radio depended mainly on height, but at around 20,000ft on UHF, the range was 100 n.m and increased to 200 n.m at 50,000ft. On VHF at around 10,000ft its range was 100 n.m. The HF radio provided Single Side-Band (SSB) or Amplitude Modulation (AM) and had 23,000 channels spaced at 1 Kc/s intervals. The notch aerial and pressurised aerial tuning units were mounted in a special panel on the leading edge of each wing root and could be switched to either aerial by the navigator. The range of the equipment in SSB was unlimited, however both AM and SSB modes of operation depended greatly upon atmospheric conditions.

The emergency UHF transmitter had two channels, one being preset and the other at a channel that would allow testing, but without interfering with the distress frequency. The set could be switched in manually or automatically depending upon the failure of the primary radio and the crew members would be informed on the central warning panels. In the event of the intercom failing there was a facility whereby the crew were able to communicate with each other using morse code. In each crew compartment was a press button with an integral lamp. If however even this failed, then as a last resort, there was a tube through which the crew could pass notes to each other in special containers!

Besides the UHF/VHF aerials in the fin tip and just below the nose, there was the HF notch aerial on top of the fuselage, just to the rear of the navigators cockpit. Other aerials included an Identify Friend or Foe (IFF) antenna, and Electronic Counter Measures (ECM). In order to maintain the maximum uninterrupted air stream over the aircraft's surface, a number of aerials were placed within the anhedral section of the wings. The idea of putting aerials in wings was not new and had actually been tried during World War 2 when, in an attempt to reduce wind resistance by removing the aerial, engineers had tried adapting the outer part of a Wellington bomber's wing and utilising it as an aerial.

**Out in the Field.**
The question of the aircraft being self sufficient out in the field has so far not been covered. This was an important requirement, made doubly so by the complexity of its electronic equipment. In the early days of sophisticated equipment such as that being used on the TSR-2, it was not renowned for its reliability by today's standard. However, it was important that such equipment was serviced and maintained to a high level, so to this end BAC were to adopt a method already in service both with the RAF and the Royal Navy. It was based on a system manufactured by Hawker

# TSR-2 PHOENIX OR FOLLY?

Siddeley known as TRACE. This was a self contained unit which was plugged into the aircraft, and by running a series of tests through the machine, the operator could establish exactly where the fault was and quickly replace the faulty component.

Equipment in the TSR-2 was, at the time, the best, but like so many new ideas in their infancy was still open to criticism, subject to high inflationary costs and vulnerable in its reliability. Many of the components used in today's modern military aircraft have their origins in the TSR-2 confirming the reasons why the aircraft should have been built. The world of computing was then very much in its infancy, and today even a common calculator has perhaps more memory than the digital computer used in the TSR-2.

**Comparison with the Tornado.**

Although the IDS version of the Tornado has the advantage of having the very latest in modern electronics a study of the aircraft's specification will demonstrate that whilst great strides have been made in this field of technology, the TSR-2 with updated avionics would still have certain advantages over the Tornado. The adoption of swing-wing technology means the Tornado can carry out certain roles slightly better than the TSR-2. The intricate and varied functions of the Tornado electronic systems, although of the next generation, are in fact very similar to those originally intended for the TSR-2. Although in certain instances the technical description of various components has changed, and the mode of operation of some of the aircraft's controls have been modified to adapt to modern warfare tactics, it is basically the same technology.

The Tornado in its ideal conditions, low down, terrain following. Only since the introduction of the GR1 versions of the Tornado has the RAF received a bomber that can be compared to the TSR-2. It is still unable to perform many of the tasks written into OR343. *(Rolls-Royce plc)*

# A MAJOR STEP FORWARD

The central computer system in the Tornado is now refered to as a Main Computer, or MC. It was originally manufactured with a 64k memory, but was soon updated to 128k and I now understand, has been further updated to 224k. Unlike TSR-2, the Tornado has only a single MC, but is supported by a number of primary sources which update the MC through a piece of software called a Kalman Filter. The filter is designed basically to eliminate rogue data before it can corrupt the MC. Equipment which helps update the MC consists of a SAHR, or Secondary Attitude Heading Reference, doppler, Inertial Platform and an Air Data Computer. The Tornado, like TSR-2, is able to perform in a back-up mode if necessary should any piece of this equipment become unservicable during a mission. This allows the aircraft to continue to function at only a slightly degraded level of performance. However, this efficiency level falls should more than one piece of equipment become unservicable, though even then its accuracy is still quite exceptional. Unlike TSR-2, the equipment takes up very little room and I am assured the standard of reliability is of a very high level.

The navigator in the Tornado works in a comfortable, spacious cockpit with the equipment falling easy to hand. Directly in front of him is the main work panel which consists of three screens and a number of instruments. In the centre is a circular screen for the CRPMD or Combined Radar and projected map display. On either side is an oblong multi-function screen, which are in fact terminals, and allow the navigator to access the MC in order to update or change the mission parameters. On the right-hand console is the control panel for the inertial navigation, doppler and SAHR.

In the TSR-2 the mission was loaded into the central computer system by a pre-punched paper-tape. In Tornado the mission is plotted onto a cassette tape from a special ground system with equipment compatible with that in the Tornado. The maps used in the ground station are usually similar to those used in the moving map display, but any map able to fit on the electronic map table can be used. By selecting any two points on the map where the longtitude and latitude are known accurately, the map can be datumed. The process of choosing a route for the mission can now begin with the navigator selecting various legs or way points, with the end of each leg being marked using an alpha character. The navigator will want to select various fix points during the mission to ensure pin point accuracy and he will choose the points, marking them with a numerical digit. The target or targets would then be entered as X, Y, or Z. Having plotted the route, information regarding the aircraft's status, such as fuel load, fuel flow, weapons carried and essential fuel for the aircraft's return will then be entered. Reliance on an accurate weather forecast is not needed as the aircraft's navigation system will cope. With all this information now on the tape it is taken out to the aircraft and loaded into the MC by the navigator inserting the tape into the cockpit voice recorder.

Alignment of the navigation equipment in the TSR-2 was expected to take around 20 minutes, but depended on the ambient temperature and the temperature of the equipment itself. In Tornado it has to be on-line much quicker, and because of modern electronics can be up and ready in around 7 minutes. To ensure the navigator is aware that his equipment is ready and functioning properly there is a multi display warning panel on the starboard console, and this will give him an indication if anything has malfunctioned. It will also draw his attention to a problem should any of the navigation equipment go unserviceable during a mission. Once the equipment is ready for use the navigator will display his mission plan on one of the two terminals, the other terminal would be used to display the

## TSR-2 PHOENIX OR FOLLY?

dynamic navigation system. The cross in the centre of the moving map display will correspond to the aircraft's actual position and as the aircraft starts to move, so the map will move accordingly. By selecting a high magnification the navigator can accurately monitor the aircraft's track. Once airborne the magnification can be altered to show a wider area. The navigator can also follow the track on the terminal with the aircraft being shown on the terminal as a small circle. The display will also show the aircraft's heading and current position. In the very rare event of there being a significant difference in the heading between the inertial navigation and the SAHR it is brought to the navigators attention by the information area on his screen flashing in reverse video. He can accept this and then synchronise the two, by bringing the I.N back into line, or instead, he can update the SAHR.

At any time the navigator can insert new coordinates anywhere into the route plan. To do this he selects the new position by manoeuvering the moving map display using a small joystick directly in front of him. This slews the map to the desired area and as soon as he can see and identify his new target or way point, by pressing the small button on the lever the new coordinates are available for use. To enter them into the route plan all he has to do now is force an interrupt in the system by changing the mode from Autosteer to Manual Hold and back into Autosteer. Any of the selected positions can be moved and inserted anywhere in the plan during the mission.

A spacious, comfortable cockpit, the navigator in the Tornado has modern equipment that can be controlled far quicker than that in the TSR-2.
*(BAe)*

## A MAJOR STEP FORWARD

The navigators second terminal will show him the aircraft's actual track, which is represented by a single line and the aircraft's planned track, shown by a pair of parallel lines which will show any cross track error. Across the top of the screen is displayed the the true heading and track. In the information area will be shown a variety of flight data including time into the mission, the time to the next turn point, and the heading from that turning point. It will also show whether the aircraft is on schedule and when the MC was last updated. At any time during the mission the pilot can take control and fly the aircraft manually.

Perhaps it is the way the equipment gathers and processes information that can easily be compared to that in the TSR-2. For instance, the Air Data Computer in the Tornado works in a similar manner to the same piece of equipment built in 1960 for the TSR-2. Using pitot and static pressures taken from various points on the aircraft's nose, the information is passed through a multiplexor before being processed in the ADC and then passed to the MC. The doppler gathers its information by transmitting three signals downwards from the aircraft's underside

*The pilot in the Tornado faces an array of instrument not a lot unlike those in the TSR-2. (BAe)*

# TSR-2 PHOENIX OR FOLLY?

A Tornado, similar to this GR1, would have carried a TIALD or Thermal Imaging Airborne Laser Designator *(Rolls-Royce plc)*

and processes the returns to give accurate speed and drift information for the MC.

As previously mentioned the Tornado is quite capable of operating effectively during those periods when it has a failure within the navigation system. The system itself can operate in one of four different modes. Firstly there is its main mode with the inertial nav, doppler, air data computer and the Kalman filter inputing to the MC. Reversionary modes operate with either the inertial nav inputting, or the doppler and SAHR or the ADC and SAHR. Each mode is capable of controlling the mission even at such a degraded level.

The Tornado employs a terrain-following radar which is an advanced, complicated system manufactured by Texas Instruments of America. This came about when the German Government put extreme pressure on Panavia to use the American radar partly to pay for the cost of stationing U.S troops in Germany. Although doing a similar task as the TFR was envisaged in the TSR-2, the Tornado TFR is a single mode system and employs a separate radar system known as ground mapping radar (GMR). The GMR has a number of different modes including ground mapping itself, terrain avoidance, air-to-ground ranging, beacon homing, height finding, and air-to-air tracking. All U.K versions of the GR-1 and GR1-A Tornado have a laser ranging system to ensure extremely accurate air-to-ground ranging and target marking. The aircraft also makes use of a TIALD, or Thermal Imaging Airborne Laser Designator pod. This sort of equipment was shown to good effect in the media during the Gulf crisis, when pictures were shown of how such a single pod could mark a target to enable attacks by the Tornado employing it and other Tornado's using laser guided weapons

# CHAPTER FOUR
## The Olympus Story

The story of the engine that was to power TSR-2 (and ultimately Concorde), begins many years ago in 1946. At the time it was rumoured that the Government was about to announce its intention to replace the present bomber force, and with this in mind, the Bristol Aeroplane Company began laying down plans of their own for a design which could possibly replace the ageing bombers. The aircraft, known as the Bristol 172, was to be a high level, long range bomber powered by four engines which would be supplied by the engine division of the Bristol Company. At the time no such engine existed, and the engine company had very little experience in jet engine design. However plans were soon drawn up for what was to become one of the world's most famous jet engines.

The normal naming convention at Bristol for new engines was to adopt a mythological Greek name, and this new engine was no exception, eventually being named after the Grecian Gods mountain abode, Olympus. The Olympus story is one in its own right, not merely for its connection with Concorde, but because it was to make significant changes away from the conventional, both with its design, and later with the innovative methods by which the engine was to be developed.

**The Olympus is Born.**
Bristol Aero Engines began studies into the possibility of producing an engine not only with a high specific output, but also with low fuel consumption, which would ideally suit the new bomber under consideration by the aircraft division at the time. It was anticipated that the 172 would need a high cruising speed of around 600 mph and that a range of 5,000 nautical miles would be an essential requirement. With these figures in mind, it was estimated that such an aircraft would require four engines, each producing 9,000lbs thrust at sea level. To gain maximum efficiency and economy, a low frontal area was important, therefore an engine with a high compression ratio axial flow system was required. This was because the engines were to be mounted in the wing root of the bomber. The company's experience in manufacturing jet engines was limited, although considerable experience had been gained with the turbo-prop Proteus.

Jet engines manufactured in that period were mainly single shaft designs, housing the single spool compressor at the front, and a single stage turbine at the rear. Furthermore, the highest output so far achieved from an engine in this configuration was only 4,500lbs thrust and it was felt that to redesign the compressor in its present form would still not produce the required results.

**Comparing the Clyde.**
In early 1944 Rolls-Royce began developing a twin shaft engine, known as RB39, the Clyde. This had a 9 stage, low-pressure, axial compressor driven by a single stage turbine. Running in parallel on an outer shaft, was a centrifugal high-pressure compressor driven by its H.P single stage turbine. It was a turbo-prop engine, and when introduced in August 1945, had an output of 2,000 s.h.p. However, after further development the output was increased to 4,200lbs s.h.p. Although quite successful, the Clyde never went into production and was only used for experimental purposes on a small number of Westland Wyverns.

Bristol's felt initially that by using a similar configuration to that of the Clyde,

# TSR-2 PHOENIX OR FOLLY?

they could overcome some of their problems. Like the Clyde engine, the Olympus was to have an axial LP compressor, but the HP compressor would use the centrifugal design. It was however felt that this configuration would lead to further complications in that the HP compressor would be far bigger than the axial LP compressor. This would cause a work split with the larger HP compressor rotating at several times the speed of the LP compressor. Ideally the rotational speeds between the two compressors should be fairly evenly matched. Bristols therefore decided to use a system where two entirely different independent axial compressors, or spools, would be connected by concentric shafts to their respective turbines. Never before had such a complex layout been used in the construction of a turbojet engine. This would give greater flexibility of engine control and superb handling characteristics.

The Bristol 172 never came to fruition, but in March 1946, the Ministry of Supply issued a requirement for an engine in the 8000lbs thrust range to power the new Avro Vulcan bomber, and in response to this Bristol submitted their new design for consideration. By July an engine specification, number TEI/46, had been issued from the Ministry and during January the following year, Bristol were given authorisation to produce six engines for experimental work. This then, was the origin of the Olympus.

*Above:* Bristol Aero Engines initial concept for the Olympus was based on the Rolls-Royce RB39 Clyde. Dr Stanley Hooker dismissed the design as unsuitable mainly because of the large differences in workload between the high and low compressor.

*Below:* A small number of the Rolls-Royce RB39 Clyde engines were used to power a few of the Westland Wyvern. The development of this engine was never taken beyond this stage.
*(Rolls-Royce plc)*

# THE OLYMPUS STORY

Double Entry Single Stage Centrifugal Compressor | Cannular Flame Tubes | Single Stage Turbine | Jet Exhaust

Induction | Compression | Ignition | Exhaust

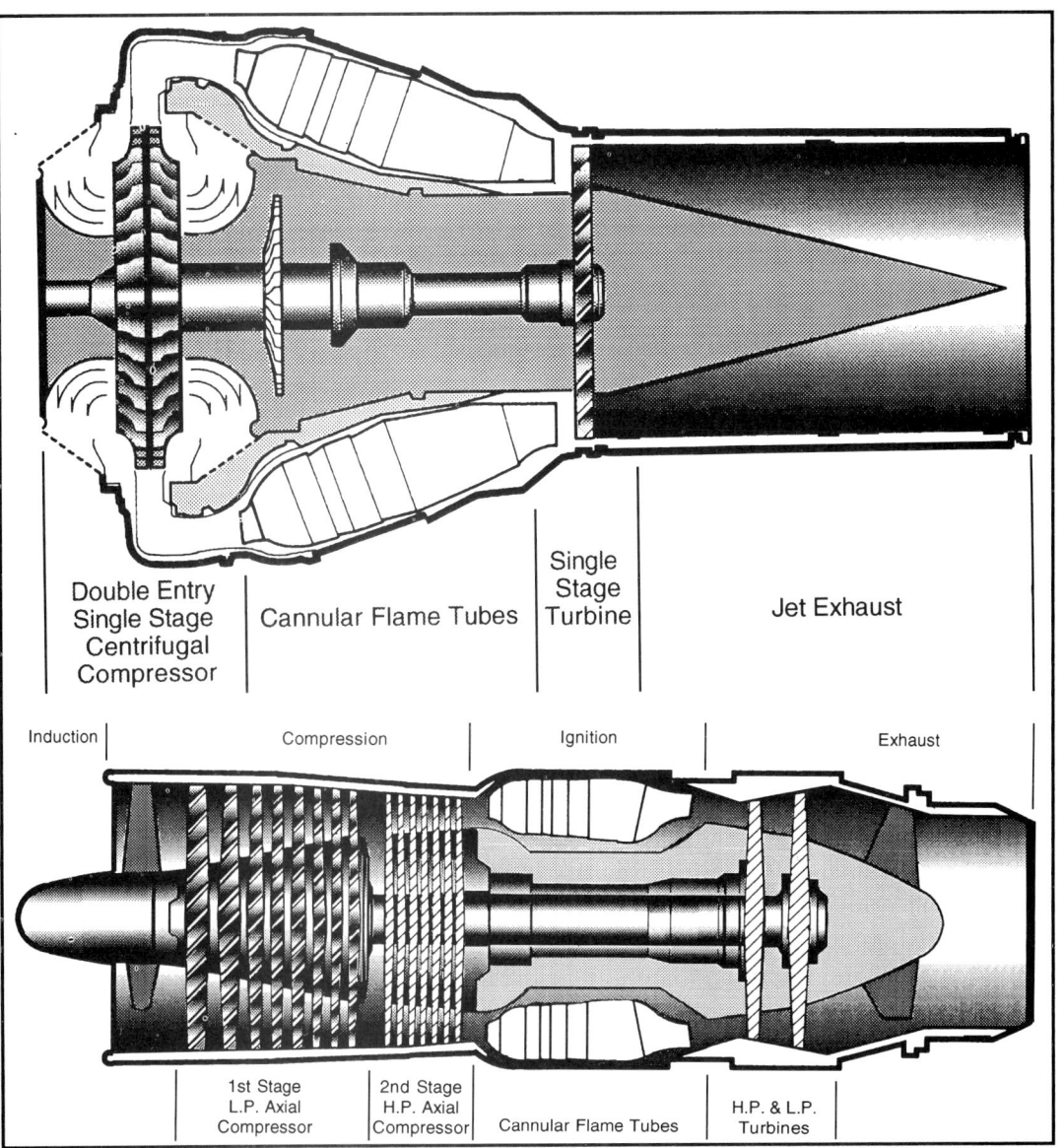

1st Stage L.P. Axial Compressor | 2nd Stage H.P. Axial Compressor | Cannular Flame Tubes | H.P. & L.P. Turbines

These diagrams show the basic differences between the centrifugal jet engine (top) and (below) a typical axial twin-spool jet engine. The working concept of the jet engine is based on the four stroke engine, Induction, Compression, Ignition, Exhaust.
*(Author Based on Rolls-Royce Diagrams)*

**Early Olympus.**

The engine had a six stage low-pressure compressor, and an eight stage high-pressure compressor, each on its own shaft driven by an individual single stage turbine. The engine, designated B 01-1, was first bench run on May 16th 1950 and produced 9,140lbs of thrust.

Sir Stanley Hooker, who at the time was Bristol's Chief Engineer, played an important role in the development of the Olympus engine and if one reads his autobiography 'Not Much of an Engineer', he describes the events when the first run of the Olympus was about to take place. If such events were as he actually describes them, then not only are they somewhat amusing, but also quite alarming. As Hooker entered the test cell, the engineers were about to cautiously start running

# TSR-2 PHOENIX OR FOLLY?

the engine. Hooker knew the process would be a slow one as the engineers would slowly run the engine in, checking every meter and noise as they coaxed the engine to run up to its eventual maximum output. So, without further ado, he took control of the throttle, started the engine and after a short period swiftly moved the throttle wide open. The engine responded immediately and within a few seconds the thrust meter was measuring nearly 10,000lbs thrust! There is some contention as to what this initial thrust was, for at the time a number of engineers could only remember a figure of 9,146lbs. However, if what Hooker says was true, he took a very brave gamble and the consequences had the engine completely failed, could have been truly horrendous, with the possible complete destruction of the test facility.

Development of the engine went ahead although not without its problems. The first flight of the Vulcan bomber was imminent and with no Olympus engine ready, Avro had to fit the Rolls-Royce RA3 Avon instead. In August 1952, the first aircraft to fly powered by the Olympus engines was a Canberra, WD952, with its two Rolls-Royce Avons removed. The Olympus engines used in this aircraft had to be derated simply because of the design of the Canberra's main wing spar. The Avon jet-pipe passed through a hole in the spar, but the Olympus jet-pipe was much larger in diameter. The hole could not be increased because of weakening the structure, so the Olympus engine had a smaller jet pipe fitted which meant the mass flow had to be reduced. Even with derated engines, had full power been used on take-off the thrust would have easily exceeded the single engine safety power of the rudder.

On May 4th 1953 WD952, powered by two Olympus B.01-1/2's, set a new altitude record of 63,668 ft. By 1955 this height was increased when on the 29th August, the same aircraft fitted with Mk 102 engines, set the new record at 65,876ft. (The Mk102 was a further development with an increased thrust of 12,000lbs). In 1956 WD952 was to meet with an accident and was subsequently written off. Meanwhile increasing development work continued on the B01-1 engine and thrust was increased to 10,000lbs, bringing a new designation, the B 01-1/2B Mk 100 and after further increases in power the Mk 101/2C was producing 11,000lbs thrust. This was the engine scheduled to enter service with the RAF in the early Vulcan B1 Bombers in July 1956.

The first Olympus engines to fly did so in Canberra WD952 and in doing so set a number of altitude records. This aircraft was to meet with an accident in 1956 and was subsequently written off.
(Rolls-Royce plc)

# THE OLYMPUS STORY

**OLYMPUS MK. 30101 E.C.U.** — INTERNAL DETAILS

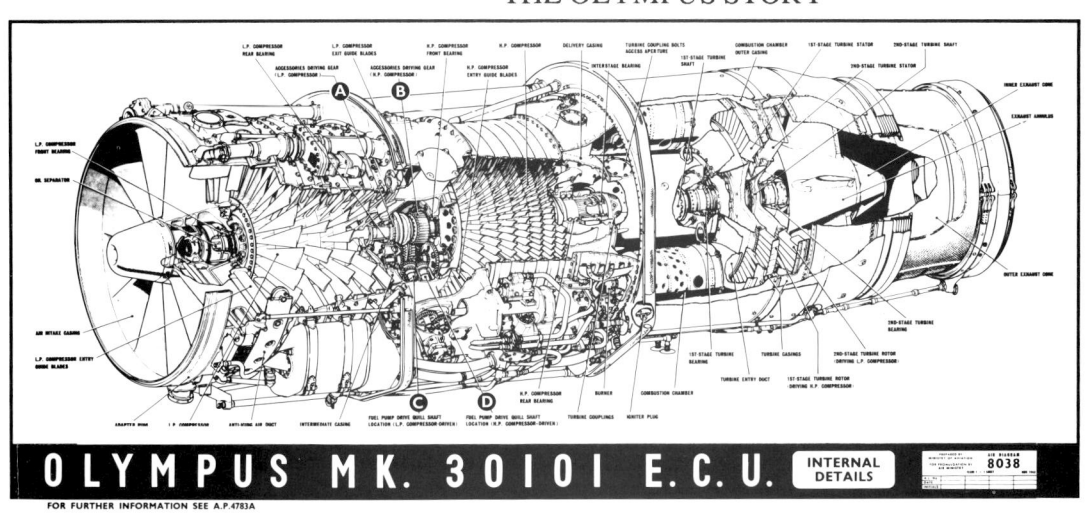

FOR FURTHER INFORMATION SEE A.P.4783A

The Olympus 301-01. The 320 engine was a derivative of this particular engine. *(Rolls-Royce plc)*

## Increased Development.

Further development of the original Olympus Mk 102, resulted in 12,000lbs thrust being achieved. This had been made possible by the addition of a further stage to the front of the low-pressure compressor. The availability of new materials made it possible for significant improvements to the compressor, allowing for greater compressor speeds and higher turbine entry temperatures. This in turn gave an increase in thrust of up to 13,500lbs, and resulted in the engine being designated the Mk 104. This new Mark of engine was fitted in the Vulcan B1's and the earlier Marks being removed and updated to 104 standard during scheduled overhauls. Confidence in the engine was now growing, especially as the development work done on the Mk 104 had brought such improvements. It was therefore decided to continue this policy of partial design to achieve even greater thrust. As a direct result of this development policy the Olympus 200 was born. The 200 series was a 12 stage, 2 spool, axial compressor, comprising of a 5-stage low-pressure and a 7-stage high-pressure compressor, with each compressor having a single stage turbine. Initial thrust was 16,000lbs but by increasing the compressor speed and the turbine entry temperature this was subsequently increased to 17,000lbs and a new designation to Mk201. The 201 was fitted into the modified Vulcan B2 bomber with its larger wing area and with its more powerful engines. It was estimated this would give the B2 not only a greater ceiling, but the ability to carry the Blue Steel stand-off bomb. With the addition of a further stage to the low-pressure compressor the output was increased to 20,000lbs, and a new designation to B 01-21, Mk 301, and the engine was again being fitted to the Vulcan B2's. By 1963 the ageing Vulcan B1's were being withdrawn from service.

In 1957 the Ministry of Supply issued the General Operational Requirement for a Canberra replacement and Bristol Aero Engines began development of an engine to meet this requirement. It was to be a derivative of the Olympus 301 and originally was to be known as the B 01-22R, but later as a production engine known as the Olympus 320. The engine was to undergo many changes in design and content to ensure it would meet the demands of its new role, particularly the stresses and strains experienced during the change from its subsonic to supersonic capability. Moreover, not only did the specification of the newly issued OR343, demand more from the aircraft, it also demanded more from the engine, particularly in better values of specific fuel consumption. One particular requirement was the need to run at full military power for 45 minutes and have an operational range of 1,000 miles. This latter point certainly did not impress Stanley Hooker, who at one time, confronted the Vice Chief of Air Staffs, Sir Geoffrey Tuttle, pointing out the penalties of insisting on such a requirement. He felt this figure of a 1,000 miles was one 'plucked out of the sky', and the enormous costs of development to attain this figure would be, he estimated, in the region of £1 million per mile over the last 100 miles. Had the Air Staff accepted the original figures given by Bristol's of between 800 and 900 miles, the costs would have been more controllable.

## Olympus 320-22R.

The 22R was a 15 stage, 2 spool, axial compressor, which comprised of an 8-stage low-pressure spool and 7-stage high-pressure spool, each compressor having a single stage turbine. The combustion used 8 cannular flame tubes with twin fuel burners. The 1st stage blades were shrouded air cooled, with the second stage being solid. The engine was designed to produce in excess of 19,600lbs thrust dry, and with reheat, of over 30,000lbs thrust, making it one of the most powerful jet engines in the Western world. The exception was the General Electric YJ93-GE-3 engine, six of which powered the XB70 Valkyrie. This engine produced 31,000lbs thrust with afterburning. One disadvantage with the G.E engine was its fuel consumption, which was considerably more than that of the Olympus. Materials used in the manufacturing of the 320 engine were also very different to the original 301 and were to significantly increase the overall weight from the 3,650lbs, of the 301, to 6,000lbs of the 320. The demands that were going to be made were far in excess of any previous Olympus engine. The aircraft would be flying 80% of its missions at Mach 2+ and the temperatures at this speed are quite excessive.

Some appreciation of temperature change over a varying speed range, may be had as follows. At 100 mph the outside skin temperature is around 1° C, at 200 mph this temperature has now increased to 4° C, and at around 1500 mph the temperature is nearly 225° C. However when flying at high altitudes the outside temperature drops to -50° C and so it was estimated that air entering the front of the engine in TSR-2 when travelling at 1,450 mph would be at some 160° C.

To cope with these extremes of temperature, a large proportion of the engine had to be manufactured in Titanium, Nimonic and similar high temperature metals. Normally this sort of material only had to be considered for use at the rear or hot part of the engine, but now it was obvious that fresh consideration had to be given for using such material in the manufacture of components throughout the entire unit. Another of the biggest changes was the addition of an afterburner, (or reheat).

In Britain the use of thrust augmentation was being used on only a few aircraft,

*At the time the west's second most powerful engine, the Olympus 320-22R.*
*(Rolls Royce plc)*

*The original reheat pipe for the Olympus was manufactured by the American Solar company. It was found unsuitable so BSEL, with the help of Solar, went about designing and manufacturing their own. (Rolls-Royce plc)*

including the Gloster Javelin and Lightning. Both Rolls-Royce and Bristol Aero Engines had been experimenting with the use of reheat since the 1950's to enhance the performance of the jet engine. Whilst this method of increasing thrust by some 30% created many problems in the development stage, the ultimate advantages and flexibility far outweighed any penalties.

## Reheat on the Olympus.

During normal jet engine operation there is a considerable amount of unburnt oxygen that passes through the engine into the exhaust. By injecting atomized fuel directly into this exhaust, combustion will occur. This creates a tremendous increase in temperature, which can be around 1,700°C, increasing the exit velocity of the exhaust gases, so extra thrust is gained. To ensure the extra rise in temperature does not affect the materials used in the construction of the jet-pipe wall, the reheat manifolds are normally mounted centrally in the exhaust. On the Olympus 320 engine, as with most other jet engines using reheat, there is what is known as a 'screech liner' which is pierced with a number of holes to provide a flow of cooling air. As the afterburner pipe is larger in diameter than the normal jet-pipe, a limited amount of pressure control has to be maintained in the jet-pipe. However, under normal flight conditions it is important that the jet-pipe orifice is maintained in order that the jet engine operates within its designed parameters. To ensure this the end of the afterburner nozzle is made variable by using either a two position movable nozzle, with two movable flaps (rather like eyelids), or, with a variable area nozzle. This has a number of flaps that can be moved inwards for normal flight, or opened when reheat is engaged.

In 1956, an Olympus engine had in fact run with a very simple American made Solar afterburner, but for a number of reasons this particular type of application was unsuitable for TSR-2. Previously, reheat had normally been used for very short periods, such as during take-off, but the particular demands made on the TSR-2 would mean the aircraft having to use reheat for much longer periods.

# TSR-2 PHOENIX OR FOLLY?

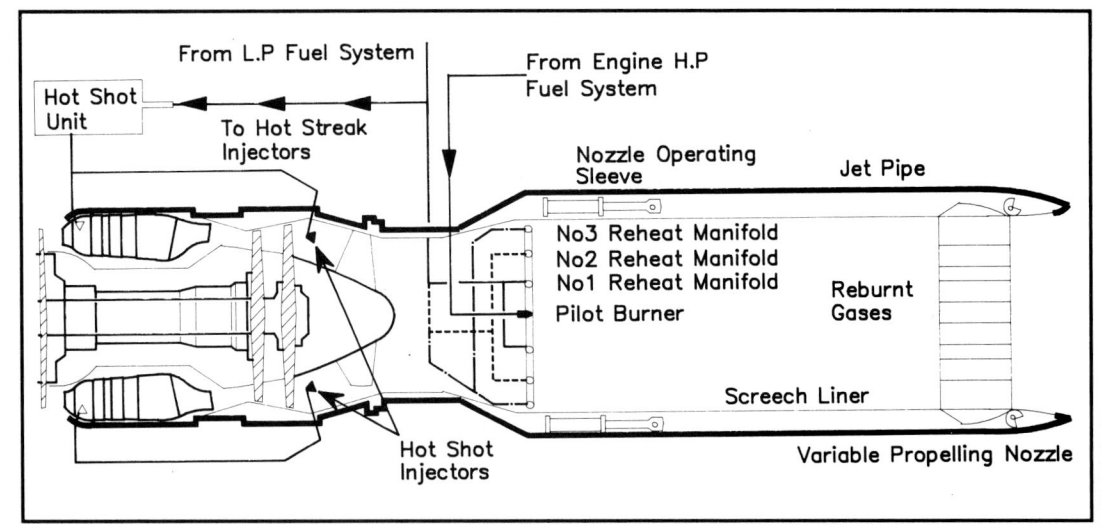

*The basic principles of reheat as used on the Olympus 320 for the TSR-2. (Author)*

During the first stages of development, the reheat type used on the Olympus was a 38 inch, hydraulically-controlled, variable nozzle jet-pipe, made by Solar. Eventually, however, it was found that the diameter would need increasing to some 40.5 inches so Bristol's, in conjunction with Solar, set about designing a suitable replacement. This was to be a convergent-divergent type, with a pneumatically controlled, 36 flap, variable nozzle. The system used three concentric rings or burner manifolds mounted centrally within the the jet-pipe. This would ensure not only an even flow, but would also, by keeping the flow centrally within the jet-pipe, help to keep the walls of the jet-pipe effectively cooler. Fuel for the reheat was taken from the low-pressure fuel system, using its own reheat pump, driven off the engine's H.P compressor. It was necessary to build certain safety features into the pump to take care of situations such as mechanical failures or excessive overheating, for should any of these incidents occur, the pump drive would automatically be disconnected and the only indication the pilot would have was the fact that reheat had been disengaged. In the event of this happening the resetting sequence could only be carried out by an engineer on the ground.

As with normal reheat selection, this was achieved by the pilot moving the throttle lever through to the maximum dry power gate, and then, lifting the latch lever to select reheat. During the very first stage of reheat selection, only the pilot burner is lit. Because the reheat manifolds were some distance from the hot turbine exhaust, the exhaust gases were of insufficient strength to ignite the reheat fuel. Therefore a form of ignition, known as 'hot streak' ignition was chosen. This involved fuel being injected at three places within the combustion chambers simultaneously. This caused a flame from the combustion chamber to be fed through to the reheat manifolds to provide ignition for the pilot burner. This 'streak' only lasted for about 1.5 seconds and would be automatically cut off after 3 seconds. The pilot burner, using fuel from the high pressure fuel system thus provided a stable form of ignition for each manifold, even in the case where one of the manifolds was extinguished during reheat running.

The pilot continued to move the throttle lever in the reheat stage and pick-up points on the lever itself relayed signals to a motor driven gearbox in the reheat selector. Stage one energized a solenoid allowing fuel to be fed, under pressure, to the No1 reheat manifold. Further progression of the throttle energized a second

# THE OLYMPUS STORY

solenoid and this allowed the fuel to be shared between the No1 and No2 manifolds. By taking the throttle to its absolute maximum the pilot would energize the final solenoid for the No3 manifold. Here the fuel was evenly distributed across all three manifolds, to give maximum reheat power. To ensure there was no pressure drop, each outer manifold had a 'fill gallery' that ensured the manifold was full before flow and ignition could take place. Because the system used an electronic method of control, the reheat selection was progressive and therefore only in extreme circumstances would it allow the pilot to select full reheat by slamming the throttles wide open and have all three manifolds burning at the same time.

The reheat system on the TSR-2 was in fact extremely reliable, unlike some systems around at the time. Perhaps understandably, BAC's Chief Ground Engineer was very sceptical of the system, having had experience on other types of aircraft in the past when the reheat had frequently failed to ignite during initial ground and flight trials. TSR-2 however, proved very effective and was to operate relatively trouble free throughout the development period, and would surprise the engineers when it did not.

## Considering the Rolls-Royce Medway.

Aircraft manufactured in the past by English Electric had usually been powered by Rolls-Royce power plants, in particular the Avon engine. It was therefore felt that the Rolls-Royce RB142, 'Medway' would be a prime candidate for the new aircraft. The Medway itself was a very similar engine to the Olympus, being a 17 stage, 2 spool compressor, producing an initial thrust of around 17,000lbs thrust dry and an expected 23,000lbs thrust, with reheat. The biggest difference was that the Medway was a by-pass or turbofan engine as opposed to the Olympus, which was a turbojet. (With a turbofan, a certain amount of the air entering the front of the engine is ducted around the central core of the engine). It was considered that the Medway was a far lighter engine and therefore would inevitably prove to be somewhat more economical than the Olympus.

The fact the Medway was a turbofan, although considered by some to be a

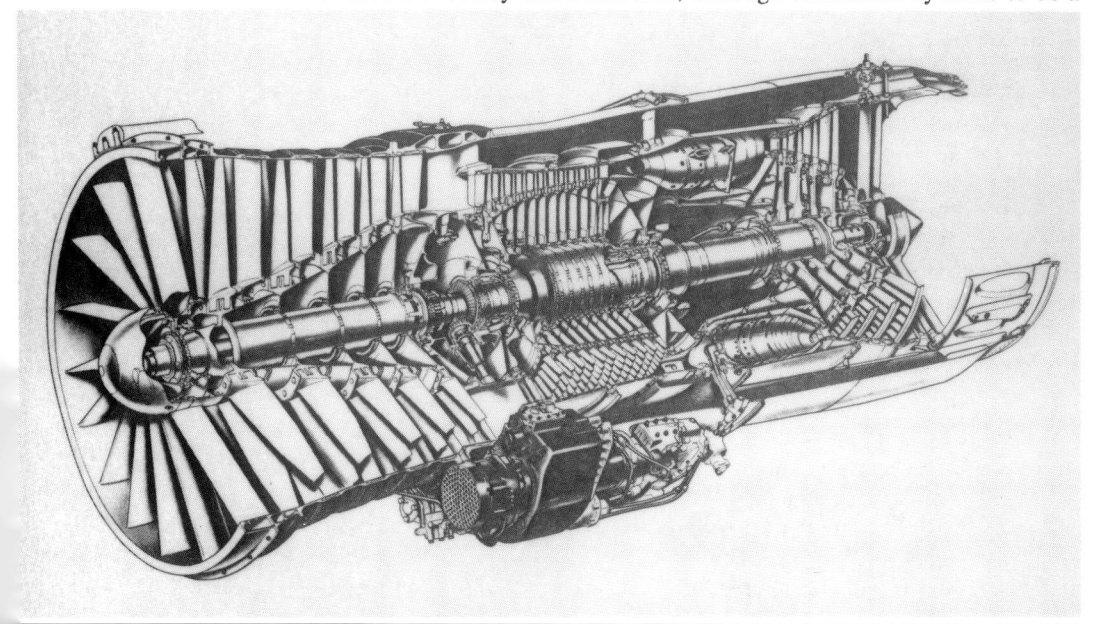

*The Rolls-Royce RB141 Medway. Considered by many as a suitable power plant for the TSR-2 the Medway was also proposed to power the HS681 transport plane. A civil version, the RB171, was also proposed for the VC10 and Trident. Designed as a low bypass engine with an initial dry thrust rating of 17,000lbs.*
*(Rolls-Royce plc)*

# TSR-2 PHOENIX OR FOLLY?

*The Rolls-Royce Medway as proposed for the HS681. The deflection plates can be seen at the rear of the engine designed to give the aircraft a STOL effect. (Rolls-Royce plc)*

problem because of extra oxygen during the reheat stage likely to cause increased temperatures, was in fact no problem at all. On a by-pass engine, the by-pass air provides a low pressure cool flow, while the core stream, that which comes from the engine, provides high pressure, so the combustion requirements are different for the two streams in the reheat pipe. At the time, the only turbofan engine that had run with reheat was the Rolls-Royce Conway, and although it had been done for demonstration purposes only, had proved quite successful. In the end however, it was found not to be such a serious problem after all, as the development of the by-pass Spey proved when that engine was fitted to the British versions of the American McDonnell Douglas F4 Phantoms. Today the majority of all modern military jet fighter aircraft now use a turbofan with augmentation.

Although figures had been produced for such a configuration, the Medway had not run with augmentation in Britain. However, in America, the Allison company had actually accomplished this and had run the Medway with reheat. The results appeared encouraging, and Rolls-Royce were confident that, with further development, the Medway could be run under the conditions required and would have soon been matching those figures produced by the 22R. The Medway had already been proposed for use in a number of applications including Hawkers HS681, the intended support vehicle for TSR-2. Sir George Edwards had also felt that the Medway was an ideal unit for the new aeroplane, and that it also possessed a far greater development potential for later engine marks. However BAC's preference was overruled by the Ministry, basically on the grounds that the Olympus was an already proven engine producing a higher thrust rating. At this stage the Medway was still not yet fully developed and was unlikely to be producing anywhere near the power required for the TSR-2 in time for its scheduled first flight, which at the time was January 1963. Consequently, in 1963, the development of the Medway was terminated with the introduction of the now famous Spey engine.

**Engine Manufacturers Merge.**
Like the aircraft industry, the Government desired some form of amalgamation between the aero engine manufacturers in an attempt to rationalise the resources being used. This would allow for strong management and better competition. A strong inducement to achieve this was the engine supply contract for TSR-2. Rolls-Royce did appear to be the natural contender, but Bristol's position required them to also bid for the contract. By 1959, Bristol Aero Engines and Armstrong Siddeley Motors had been forced to amalgamate and in 1961 were joined by the engine divisions of DeHavillands and Blackburns, with the airframe division of these

# THE OLYMPUS STORY

companies going to Hawkers to help form the Hawker Siddeley group. Now known as Bristol Siddeley Engines Ltd, (BSEL), the company had demonstrated their ability to conform with Government thinking and had become a powerful company to contend with. Therefore, they were duly awarded the contract for the supply of engines for TSR-2.

**Tentative Testing.**
In March 1961, the Olympus 22R was testbed run for the first time, but its performance was well below the required standard. Late in 1962, during tests, a problem occurred involving the specially cast turbine blades. Subsequent investigation showed that one of the blades had broken up, causing the rest of the blades on the disk to disintegrate also. It was therefore concluded that the blades were of too brittle a quality, and as a result, it was decided to change the form of construction and instead of casting the blades, they were forged. As an offshoot of this it was found that by changing the blade construction in this way, fuel consumption was also improved.

The Olympus was the first engine in Britain to use a twin axial spool configuration, but the 22R derivative was the first Olympus to fly supersonic, which, although a notable achievment, did prove very costly. This configuration, as explained earlier, had the low-compressor running on the inner shaft and the high-pressure compressor running on the outer. These shafts run the entire length of the engine with each compressor having its individual turbine stage at the rear. They also run at slightly different speeds, which allows the flexibility in the engine and because of the high temperatures generated in the engine, air has to be circulated in order to cool both the shafts and the bearings. In the early Olympus engines there were three bearings. The front bearing was cooled by air from the intake, and the centre and rear bearings being cooled by air bled through special air ducts. With the aircraft flying at supersonic speeds, the air entering the engine would be some 160° C, coupled with heat generated by the engine itself, Bristol decided it would be better to eliminate the centre bearing, and so lessen the risk of fire in this area. To ensure the shaft remained stable, the manufacturing process was modified to increase its diameter.

**Electronic Engine Management.**
The Olympus was also the first military engine to use a simple electronically controlled throttle system, the forerunner of fly-by-wire. This method of controlling the throttle on an engine was not new, and had been pioneered in 1946 when a crude method of electrical management was devised for use on the controversial Bristol Brabazon airliner.

It became apparent that with the size of aircraft increasing at the present rate, the current method of engine control was going to become unmanageable. Sheer size alone was bringing problems, not only with the need for more powerful engines, but also the necessity to move away from the current system of control with its rods, cables, pulleys and levers. The new and complex designs were not compatible with the old method of control and, with new speeds and heights anticipated in airline travel, the increasing stress on airframes would ultimately affect the management of vital controls.

The Brabazon, with its 8 Centaurus XXs engines connected to four sets of contra-rotating propellers, used a magnesyn positioning system for throttle control. The 400 hours of development flying the aircraft did before it was

## TSR-2 PHOENIX OR FOLLY?

*The Bristol Brabazon with its eight Centaurus XX's was one of the first instances where a form of electrical management was used for engine control. The Bristol Freighter in the background is flying 'chase'.
(Rolls-Royce plc)*

scrapped, was sufficient to prove that such a system could quite easily work and with more development had great potential.

In 1949 Saunders-Roe launched their enormous flying boat, the Princess. This was powered by 10 Proteus turboprop engines coupled together, but, because of its size, and the complex route a normal linkage would have to take in order to reach the engines, the designers once again turned to electrical power to control the engines. This powered the control for acceleration, compressor speed, and served a jet-pipe temperature limiter. With the addition of a few more controls, results showed that it too was an easily workable system with many advantages over the hydro-mechanical system. By eliminating parts the need for servicing and attention would be less frequent, avoiding consuming delays and the aircraft would spend less time on the slipway. The major components that managed the system could be mounted in more accessible places and sited well away from hot engines that tended to vibrate components excessively. The saving in weight would be considerable, but most important, it would give better response to engine control.

*After the Brabazon was the Princess flying boat with its complex engine layout. By now confidence had grown in electronic management and the Proteus engines on the Princess used a more sophisticated management system.
(Rolls-Royce plc)*

# THE OLYMPUS STORY

With the Britannia came the ultimate in electronic management. As a long distance passenger aircraft reliability was important.

*(Rolls-Royce plc)*

During the early fifties the Aircraft Division of Bristols introduced its Britannia, nicknamed the 'Whispering Giant'. Powered by an advanced version of the Proteus, it too had a modified electronic system. When it entered service, in December 1955 with BOAC, it was the first passenger carrying aircraft to use an electronic engine management system. As the Britannia had been designed as a long distance aircraft, the reliability factor was paramount, and power for the system was provided by the aircraft's main electrical system, which was 115v AC, single phase, 400 cycle, with a manually controlled reversionary back-up method, using a 28 volt DC supply. If the main throttle system failed, reversion to manual was achieved by an 'inching' switch situated in the cockpit. The DC motor, which drove the throttle, had to be used with caution, as on one occasion over zealous use caused an engine on a BOAC airliner to burn out! The chances of both systems becoming inoperative at the same time were considered highly improbable.

This method of engine control seemed the natural way forward when consideration was being given to the specification for the Olympus in TSR-2. Here was an aircraft that was not only to fly in excess of Mach 2 but also fly skirting the contours of the earth where turbulence would induce tremendous buffeting and vibration. These exacting requirements laid down by the Operational Requirement more or less demanded the need for electronic management. The engines in TSR-2 were side-by-side in the rear of the fuselage, housed in separate tunnels where space around the engine was very restricted and subject to excesses of heat, especially in the supersonic period. Designing a method of control isolated from such temperature extremes would certainly be an advantage.

**The 320 Takes to the Air.**
With some 1900 hours of simulated running time on the test beds at Bristol and the National Gas Turbine Establishment at Pyestock, the Olympus 22R had almost reached the stage when it was ready for a flight test. In 1960 the Ministry of Aviation supplied Bristol with a Vulcan, XA894, to allow the 22R engine to be air tested. The engine was attached to the belly of the Vulcan with the fuel tank for the 22R fitted in the bomb bay. The 22R had a bifurcated inlet at a position below and slightly to the rear of the Vulcan's own two inlets with a single exhaust terminating

## TSR-2 PHOENIX OR FOLLY?

at a similar position to that of the Vulcan's four exhausts. Underneath the rear of the fuselage, was an acoustic shield that housed a periscope, and this provided an attachment for a 35mm camera to film the reheat during operation. This particular piece of equipment was never successful and nothing was gained from its use. On February 23rd 1962 the Vulcan took to the air with the 22R. This was the first of some 35 flights. The performance of the 22R exceeded all expectations, and the flight crew were able to throttle the Vulcan's Olympus 101s right back and actually fly on the 22R, although the Vulcans own engines did have to be left running to supply electrics and hydraulics. The air testing progressed satisfactorily and the Vulcan went on to appear at that year's Farnborough Air Show. The 22R's variable reheat up until this point had been tested using the first two stages only, and therefore it was essential that a test using all three manifolds be scheduled. On November 12th, this test took place and appeared to go quite normally. On board was BAC test pilot Roland Beamont who took the controls of the Vulcan in order to assess the engine handling and reheat characteristics.

The 320 takes to the air. Between the two Vulcan's are four generations of Olympus engines. (Above:) The lead Vulcan B2 XA997, powered by two 201's on the starboard side and two 301's on the port. XA894 (below), the flying testbed for the 320, is powered by four 101's with the 320 under the belly. The acoustic shield with the periscope can be seen clearly here at the tail of the aircraft.
*(Rolls-Royce plc)*

# THE OLYMPUS STORY

**Disaster on the Ground.**

It was on February 3rd 1964 that disaster struck. The engine had been scheduled for a ground power run, and on the chosen day the crew on board the Vulcan consisted of John Cruse, (Pilot), Chris Haythornthwaite, (Air Electronics Officer), John Dickenson, (Flight Test Observer) and Richard Elder, (Flight Test Observer). The engine exhaust was fed into a detuner which was water cooled during the reheat stage of the test. The ground run was going normally and so it was decided to select the reheat 3rd stage. In the words of the Flight Test Observer, 'There was suddenly a tremendous thump and bang', and the aircraft was physically moved forward under the force of an explosion. The pilot shouted, 'All four fire warnings, everybody out'. Richard Elder was responsible for getting the escape door open, and having done so, he jumped out, virtually surrounded by burning fuel, quickly followed by the rest of the crew, all running for their lives. The crew had practiced emergency escapes from the aircraft in case of accidents and so far they had achieved a time of 45 seconds to get all the crew out. In this instance they estimate it took them just 32 seconds to get out and clear!

It later emerged that there had been a catastrophic failure in the Olympus engine, the first of four which were to occur before first flight. The low-pressure drive shaft had failed having allowed its associated turbine to overspeed and become uncontained. The L.P turbine disk had then shot up into the Vulcan's bomb bay with a tremendous force and then crashed downwards, bounced off the ground, and smashed through the aircraft's leading edge on the port wing, before bouncing a further 14 times and coming to rest just 50 yards from the Bristol 188, all stainless steel prototype aircraft which was sitting at the end of the runway. The actual turbine disk had scythed its way around the aircraft's bomb bay, and had sliced open

*Disaster on the ground. This dramatic shot of XA894 shortly after the uncontained failure. All that was left of the Vulcan at the end of the day was the outer portion of the port wing.*
*(Rolls-Royce plc)*

## TSR-2 PHOENIX OR FOLLY?

the test engines fuel tanks which had been mounted in the bay. When the disk hit the ground turbine blades shot off like bullets and penetrated the Vulcan's own fuel tanks. The fuel quickly ignited, rapidly spreading down the sloping hard standing area toward a fire engine that had been on standby during the engine runs. Bristols had only recently acquired this vehicle, but it was quickly engulfed in flames and totally destroyed. The aircraft also was a total write-off, and all that remained of the Vulcan at the end of the day was the outer section of the port wing. This incident was to be one of the most expensive aviation accidents that year.

**A Serious Setback.**
An investigation as to why a turbine disk should become detached in this manner was launched immediately, which continued well into the development programme and was not resolved at the time of the first flight of TSR-2. A replacement aircraft was sought in order to continue the air tests, and one was actually identified. Although much would have been gained from continuing air testing the engine, the proposal was turned down, the deciding factor being the amount of time it would have taken to convert the bomber. The conversion of XA894 had taken nearly two years and by which time it was hoped the TSR-2 would have flown. This indeed was a severe blow to Bristol Siddeley. The development costs were increasing and now, to make matters worse, the engine had failed in an aircraft. Bristol had originally quoted some £20 million for the research programme. This greatly exceeded the Ministry's own estimate of £12 million and therefore Bristols were obliged to review their research programme costs. This was to put even further pressure on the engine programme.

**Further Setbacks.**
During the investigation into the failure it was found that the low-pressure shaft had fractured, allowing the turbine disk to become detached. At the time the cause of the fracture was not fully understood, there was however, an attempt to overcome the problem, by doubling the thickness of the metal on the shaft. The Olympus subsequently performed satisfactorily for a considerable number of hours on test beds and Bristol began thinking the problem had been cured. This however, was somewhat premature because in July 1964, just before the first flight of TSR-2, a second failure of a similar nature occurred.

This second failure happened at the company's test beds at Patchway, Bristol, and the circumstances were very similar to those in the Vulcan. Although this time the L.P turbine disk was retained within the engine, the H.P turbine disk had

Ground runs begin with TSR-2 and the Olympus. During high speed ground runs the Olympus engine needed all the air it could get. This shot shows the ground running auxillary doors open just in front of the serial number. *(BAe)*

# THE OLYMPUS STORY

*The Olympus 320 was a very powerful engine, but it was also very dirty as this shot of XR219 shows during one of the early ground runs. (Rolls-Royce plc)*

become detached, exiting the engine and wreaking havoc in the test cell. The disk, with its turbine blades still attached, became in effect a huge circular saw and had actually circled the test cell a number of times cutting and slashing anything in its path, including some of the building itself.

A General Election was looming and with the odds being on a Labour Government being elected to power, the pressure was building up on the manufacturers to get the TSR-2 into the air as soon as possible The engine problem had still not been properly identified, but Roland Beamont after discussion with the Controller Aircraft, the English Electric and Vicker's management and engineers and designers at Bristol, very bravely decided to go ahead with a test flight in TSR-2. The first aircraft, would use engines with some estimated 5 hours flying time left in them, though even this was not guaranteed by Bristol. Even then it was estimated that the L.P. shaft catastrophic failure was likely to occur at anytime when 97% was exceeded. These engines were designated 320X indicating that they were of an early build type and that the turbines were of lower efficiency. As it had been discovered that 'resonance' could be heard a few seconds prior to an engine failure, a warning system was fitted following Flight 1. It consisted of two warning lights in the pilot's cockpit, one for each engine. The lights were driven by an amplified output from a strain gauge fitted to the LP shaft. This was a fraught procedure as strain gauges at low pressure are somewhat notoriously noisy, and as such the warning light would not distinguish between amplified noise or genuine stress. The lights were supposed to warn the pilot that an engine was in imminent danger of exploding, and would also hopefully give the crew sufficient warning for them to vacate the aircraft. Roland Beamont says in his book, Phoenix into Ashes that he had been very sceptical of this modification as it was itself prone to single system failure and indeed he was proved right, for, during Flight 3 at the critical take-off stage, both red lights had flashed their warning. He was then faced with a vital decision. Should he and his navigator abandon the aircraft, thereby putting the development of the TSR-2 project back many months and perhaps even worse, jeopardising the whole project there and then, or should he continue with the flight and hope that it was no more than a false alarm, as it was certainly improbable that both engines were about to fail. It was indeed a false alarm, but it was only Beamont's skill as a test pilot together with his calculated decision that averted a major disaster in the development programme. This warning system was subsequently removed to stop similar false alarms.

**The Problem is Discovered.**
From the time of the second failure, it was not until the autumn of 1964 that the problem was fully identified and a modification programe initiated. The National Gas Turbine Establishment at Pyestock, working in conjunction with Bristols, finally identified the cause. It appears it was the reduction to 2 bearings on the L.P. shaft that resulted in an unstable cooling flow setting off the shaft resonance. It was found that air, bled into the low-pressure shaft through a diaphragm between the shafts was causing resonance, by 'exciting' the low-pressure shaft. The effect was identical to that of a bell ringing. As a bell deforms when ringing, so was the LP shaft when resonating to the extent that the shaft was eventually destroyed. This happened at the top end of the power range above 97% exactly at the critical period when the aircraft would be in the take-off stage. Bristol immediately started a temporary modification programme. Two rings were heated and shrunk round the shafts in order to alter the resonance and secondly, the air bled into the shafts was cut off. Although the problem had been found and a temporary fix made, it took nearly a year to modify the engine to production standard, by which time the project had been earmarked for cancellation. Modified engines to the temporary standard were fitted for Flight 2 onward and gave no further trouble.

Not all problems related to the engine were generated by failures within the power plant itself; it was found that when a number of the units were shipped to Weybridge it was impossible to fit them in the airframe due to the large amount of ancillary equipment encroaching into the fitting area and taking up valuable engine space. This necessitated a quick rethink and redesign of the area.

**Engine Ground Runs in TSR-2.**
By March 1964 XR219 had been dismantled at Weybridge and shipped in parts to Boscombe Down where it was rebuilt in preparation for its maiden flight. However, before this flight there would still be many hours of testing on both engines and equipment. The pre-flight testing itself would also give the media its first sight of the TSR-2. These runs were done on the far side of the short runway at Boscombe and across the road two haystacks provided the press with a grandstand view of the preparation, testing and engine runs throughout that beautiful summer. The early engine ground runs were carried out by Bristol Siddeley Senior Service Engineers,

# THE OLYMPUS STORY

XR219 at Boscombe Down undergoing engine trials. This rare picture shows the noseleg extended on TSR-2. Although designed to give TSR-2 assistance at take-off, it was found not to be necessary because of the power of the Olympus engine. It did have one useful purpose, after the 'wet' engine runs the noseleg was extended to drain off excess fuel. (BAe)

which included Phil Pearce and by one man in particular, Richard Elder, the engineer who had made the miraculous escape from the Vulcan bomber after the uncontained failure. His job was now done from the back seat of XR219 and he can remember sitting in the back with his only means of escape being simply an axe to smash the cockpit and a length of rope to climb down from the aircraft. The mind boggles in trying to visualise him wielding it in the confines of the rear cockpit had an emergency arisen requiring him to evacuate the aircraft in a hurry.

A series of problems meant that it was well into April before the Olympus engines began running properly. The initial checks on the engine were done 'dry', that is by using the engine's starter motor to turn the engine over, without fuel or ignition, until it reached the required speed where it would normally start. The first time this was attempted the engine only managed to attain half the required speed, and it was discovered that a cross feed valve had stuck in the open position, consequently causing both engines to revolve. The next test that was done is described as a 'wet run'. Again the engine is run up to its normal starting speed on its starter motor and then fuel is fed into the engine as normal, but without igniting it. This actually serves two purposes, one to ensure the integrity of the fuel system itself and secondly, to establish the integrity of the draining mechanism of escaping fuel. After each 'wet run' the noseleg of the aircraft was extended in an attempt to ensure any fuel left in the fuselage was fully drained off.

**Testing Continues.**
The Bristol Cumulus was not fitted to XR219 or XR220, and the engines were started by a 'Palouste' air starter through a Constant Speed Drive Shaft (CSDS). Although used primarily for starting the engines, it also served to supply a constant feed back into the aircraft for the electrical supply and air-conditioning. A system of valves within the aircraft would often stick, causing the Palouste to backfire. This was overcome by using some very old-fashioned technology indeed, the good old wooden mallet! The engineers would hit the various valves thus encouraging their operation. On one occasion the valve in the constant speed drive shattered and pieces went back into the Palouste causing severe damage. As this happened late at

# TSR-2 PHOENIX OR FOLLY?

*Preparing an Olympus 320 for delivery to Weybridge. These specially designed cradles could be rotated so that any part of the engine could be worked on with relative ease. (Rolls-Royce plc)*

night, Bristol Engineers had to run round there and then looking for spares to repair the Palouste. There were to be many more problems, and testing and ground runs went on well into the warm summer nights. Some of the engineers, despite having comfortable rooms at the local inns around Boscombe Down, were sleeping in the tents, ( originally erected for meals to be served in), so as to be near at hand as and when problems occurred. Bill Barret, another of the Bristol Field Engineers, told me the food was good and always freely available, invariably though it was cold when they got it. One particular favourite was a certain individual fruit pie, and these were consumed by the hundreds, (increasing waistlines daily), but the work never ceased in the round-the-clock efforts to sort the problems out. Staff members who were not paid for working overtime, were often there till late at night, even working long weekends. They remember also that year, very few who were on the TSR-2 project managed to take the Bank Holiday.

One particular incident occured after a modification to some wiring, necessary to correct a separate problem not connected with the engines. During the next engine run using reheat there was a noise rather like a huge pneumatic drill. Apparently it was so loud, it had been heard some distance away. The engine was quickly throttled back whilst an inspection took place. At first it was thought that the infamous 'resonance' problem was back. The external inspection proved negative. All the test instrumentation in the nearby caravan was reset to measure tortional stress and the engine was again run up with reheat selected. The same phenomena occured again with the drill like noise, but as it was only during the reheat stage checks were possible to ensure no physical damage had been done.

The reheat is served by three rings in the jet-pipe and the amount of fuel supplied to each ring is controlled by a fill gallery. Each gallery is activated by a solenoid which in turn is controlled by the throttle. The more throttle, the more rings issue fuel. The problem was in fact narrowed down to the outer manifold fill valve failing

# THE OLYMPUS STORY

to open. The solenoid which activated the valve had been inadvertently severed during the earlier electrical modifications, and because of this, the normal fuel supply for three rings was being forced down just the two remaining inner rings. This caused a severe local overfuelling, the result was combustion instability making the sound like a pneumatic drill. After the test on that particular engine, it was stripped down and inspected. The L.P shaft was found to be bent with a curvature of .012 inches.

Another problem that was to affect the engine runs was caused by the main reheat pump. This was a vapour core device manufactured by Dowty. Fuel was fed into the centre of a centrifugal impeller revolving at high speed. When reheat was not in use the impeller ran dry. This meant that when reheat was engaged the impeller had to be primed. To achieve this the impeller was run on stall, or at its maximum delivery pressure, which was around 2,000 lbs per sq inch. Once priming had been achieved the fuel pressure would return to normal. However occasionally the pump would stay on stall which would affect the reheat fuel values. Should reheat be selected on both engines and one pump stick on stall, the imbalance in fuel consumption would likely effect the trim of the aircraft.

**Controlling the Output.**

Fly-by-Wire was still very much in its infancy and it was a brave step by Bristol Siddeley that they elected to implement this development on a project such as TSR-2, that was already so full of new innovations. Although the engine designers had every choice available to them, it was felt this method was the way forward, particularly when one considers the airframe design and the roles demanded of the aircraft. Though the mechanics were different, the basic layout of the cockpit

Changing the engine on one of the RAF's Tornados, a somewhat easier task than that of the TSR-2. The operational requirement for the Eurofighter insists the same job be done in less than 45 minutes.
*(Rolls-Royce plc)*

looked the same. Each engine had its own throttle control lever, this being situated on the port console in the pilot's cockpit. Each lever controlled several functions besides throttle control. Both levers had engine start/relight buttons, override throttle inching switches and latch levers, however the No 2 or inner lever, had the radio transmit button and air-brake, whilst the outer lever, the No1, housed the radar ranging/T.F Mode switch. The two levers could be moved independently of each other or together with the Latch lever being operated to progress through each of the various stages. It was this particular function that gave Roland Beamont problems during the early ground runs. When increasing power the throttle lever is moved through the first Latch gate from the H.P cock-off to cock-on/idle, and after that there is no further latch gate right through to maxium-dry. It is then operated again to engage reheat. When reducing engine power, the first lift gate is engaged when coming out of reheat and then again to move out of idle and into cock-on. It was at this stage when Beamont accidently closed the throttle into H.P cock-off, which cut the engine, leaving the aircraft with no electrical power.

The throttle lever movement was an AC powered system and had pick-off points and micro-switches that regulated the fuel flow to each engine through its respective Throttle Control Unit, or TCU. Each lever had a preset load of 3lbs over the full range to give the pilot 'feel'. The TCU for each lever was situated in the main equipment bay, and it was this device that was the memory for all the instructions that controlled the engine. The TCU was programmed to control fuel flow in relation to the throttle lever position, with the system being designed to make it impossible for the pilot to make any abnormal demands on the engine by giving the TCU a complex set of rules. Whilst these rules may have varied slightly depending on the attitude of the aircraft, certain functions within the engine had to be maintained to ensure a smooth and continuous operation. Maximum permissible acceleration was given depending on a number of variables conforming to a set standard. Basically, it was dependent on the high-pressure Turbine speed, the total intake pressure and total intake temperature. Therefore, when the pilot moved his throttle levers, the throttle valve moved accordingly, although, had the pilot made an excessive demand on the engine by slamming the throttles wide open and exceed the set values, the throttle valve opening was reduced or stopped. This was irrespective of the position of the lever and also applied if the lever was moved into the reheat stage whilst the engine was still accelerating. Should any of those stages be overruled, an extra safeguard had been built into the system and this was based on the temperature rise in the Turbine Interstage. Should the temperature rise more than $6°$ of its datum point then the throttle valve opening was altered accordingly and if the change was excessive the valve could even be reversed.

Should the AC electronic system fail in either or both throttles, the DC throttle standby was initiated by the pilot who selected the appropriate throttle emergency switch on his port console in front of the throttle levers. He then controlled the throttle by the 'inching switch' on the throttle levers. Operation in the back-up DC mode offered a limited service and basically operated the throttle valve only. The last values used were held including any reheat, although as soon as the reheat was cancelled it could no longer be reselected. Other automatic controls lost were the H.P. and L.P. compressor, maximum speed electrical governor, the Turbine Interstage temperature limiter, the compressor delivery pressure limiter, acceleration control and double datum idling. It was essential the pilot used extreme care when flying the aircraft in this mode, handling the engine with caution and being constantly aware of instrumentation.

# THE OLYMPUS STORY

The F-111 showing the boundary layer splitter plate, compared to the TSR-2 plate in the next picture. Although considerably redesigned, the F-111 still experiences problems with compressor surge.
*(Ian Frimston)*

## Controlling the Air Flow.

The intake on the TSR-2 was designed by BAC to conform to a set of entry conditions required at the engine face as laid down by Bristol Siddeley Engines Limited. On Concorde there is a complex set of doors to ensure optimum intake conditions over the whole speed range. On TSR-2 this was handled by the half cone centrebodies in the engine intakes. The automatic operation relied upon the aircraft's speed and the temperature of the air entering the intake. Using these calculations, a mechanical hydraulic-sensing unit altered the shape of the cone accordingly. It was not until around 0.9M was reached that any great difference in

The TSR-2 splitter plate which vented the air up and over, or down and away. Boundary layer air was not a particular problem, mainly do to the long fuselage before the intake.

*(Ian Frimston)*

the cone shape was noticed. Up until that point the cone was kept to a minimum to ensure a maximum intake area for subsonic flight. There was a manual override system in case of hydraulic failure that would be operated through the inching switch on the throttle lever.

**Water Injection.**
The Olympus was also one of the first jet engines to make use of water injection. This was designed for use mainly in hot climates in an effort to improve engine performance during take-off where such performance would be greatly impaired by the high ambient temperatures. The water was held in a tank located between the engine tunnels and the process was initiated by the pilot moving a switch on his Rapid Take-off panel. When enabled, the operation was automatic and depended on the high-pressure compressor speed reaching 80%. Firstly the air shut-off valve opened and the air driven turbine and pump were run up to their speed with the water pressure opening the shut-off cock. This resulted in water being injected into the combustion chambers. Provided the tanks were full, the whole process would only last 45 seconds before the tank was dry. When the control switch was moved to its opposite position any residue left in the tank was drained off. Concern, should this last action fail to occur, necessitated the fitting of heating elements in the shut-off cocks as these were liable to freezing during a high altitude sortie. Today's turbo-jet engines use a water methanol mixture to boost power at take-off and also to assist in maintaining power at the greater altitudes where the atmospheric pressure decreases and also where the ambient temperature increases.

**Servicing in the Field.**
As with power plants on all modern bombers and fighters, access should be easy in order that servicing and fault diagnostics can be carried out quickly and efficiently, especially if the aircraft is in the front line in time of conflict. In the majority of military aircraft today, the engines are buried deep within the fuselage, in a deliberate attempt to keep the drag factor very low and produce an aerodynamically clean aircraft. Problems here arise when the time comes to remove the engine for routine inspection and maintenance, or specific changes due to mechanical problems. It appears at least some consideration had been given to this problem in TSR-2, for the OR insisted on an easily maintainable aircraft, not just for engine maintanence, but for all the other equipment carried. However, when the designers were working on the engine bay design, they came up with a novel idea. The Olympus engines and their jet-pipes were mounted on a rail system that allowed the units to be withdrawn from the airframe with relative ease, although, as with the Lightning, the engine could not be removed without first withdrawing the reheat pipe. When studying the maintenance manual for this particular operation, it appears quite complex, even though special equipment had been built to help with the operation. The jet-pipe had been designed to be withdrawn as a separate unit and had its own trolley jack, as indeed did the engine. The jacks had their own built in rail that had to be aligned with its counterpart in the airframe, which ran along the engine bay. Before engine or jet-pipe could be removed, the aircraft had to be jacked up until the main undercarriage wheels were approximately 12 inches off the ground. This was to ensure that there was no movement in the airframe when either the engine and jet-pipe were withdrawn. After the tail cone was removed, the jet-pipe trolley was wheeled to the rear of the aircraft and lined up with the corresponding rail. The whole jet-pipe could then be slid out. The pipe could then

# THE OLYMPUS STORY

be lowered on the trolley and moved away for overhaul or repair. In order to remove the engine itself, virtually all its ancillary equipment had to be stripped off whilst it was still within the airframe. Then, having released the engine locating mountings, the engine unit could be withdrawn using the same method as that employed in the removal of the jet-pipe, and again finally loading the unit directly onto its trolley.

In this way the engine could be moved easily to an area for repair or overhaul. Although perhaps slightly easier and quicker than the Lightning, it was estimated a clear days work would be required to change a single engine unit. It may be of interest to note that engine maintenance has progressed tremendously over the past decade. As an indication of this both engines in Eurofighter have to be changed within a 45 minute period. It is thought that 40 minutes is the quickest so far.

**In Case of Fire.**
Each of the TSR-2's engines had its own fire detection and suppression system. Fire detection was by firewire elements that circled the engine and jet-pipe in those areas where fire was a greater risk. The fire suppression system was manufactured by Graviner and used Methyl-Bromide as its agent, sprayed through dual head valves. A system was devised whereby, in order that the fire suppression criteria could be met, each engine was individually protected and divided into three zones. Zone 1 was by far the largest providing cover for the engine and accessory bay. Zone 2 took the form of a firewall between the engine and airframe. Zone 3 provided fire suppression in the area of the jet pipe. Each Zone 1 was interconnected so that a fire in a Zone 1 area could be backed up by the adjacent system.

**Looking at the American Engine.**
In the previous chapter, reference has been made to the proposed successor to TSR-2, the TFX, or F-111 as it later became known. Here, I intend to make some comparison between the Olympus engine and the one that was to power the F-111, the Pratt and Whitney JTF10A-20/TF30, and some idea is given as to the basic differences between the two engines. The Pratt and Whitney engine was first conceived as a private venture in 1958. The TF30, differed from the Olympus in a number of ways. Unlike the Olympus, the TF30 was a low-bypass engine with a three stage front fan. It was however, a twin spool engine having a 6 stage LP compressor and 7 stage HP compressor. At the hot end, the turbines were slightly different in that the LP turbine had 3 stages, with the HP turbine, like the Olympus, having only one. Power output from a single, fully augmented, TF30 in the early days was 18,500lbs. Two of these powered an F-111 weighing in at around 50,000lbs. Later versions of the TF30 had increased power output and today are generally in the region of 20,000lbs thrust. (TSR-2 weighed around 76,000lbs with 2 Olympus engines producing a combined thrust of 60,000lbs +, augmented, at take-off). The TF30, though, was to give the F-111 a number of problems. Although various other manufacturers had submitted tenders for the contract to supply engines for the F-111, it was the Pratt and Whitney TF30 that was chosen. One particular company that had tendered was Allison Aero Engines, whose submission was based on an engine being built in collaboration with Rolls-Royce. It was perhaps this collaboration factor that led to the engine being dismissed as a contender. Designated the Allison AR168, the unit was based on the Rolls-Royce RB168 Spey and at the time, was considered by many as an ideal unit for the F-111.

However, because of the engine's origin being with a so called foreign power, the Selection Committee discounted it. The TF30 did, in fact, go on to acquire something of a reputation as an 'unreliable' engine, especially with the US Navy, who even tried to obtain an alternative power plant for their TF30 engined F14 Tomcats. The major problem with the F-111 though, did not appear to be within the engine itself, but in the actual design of the aircraft's engine intakes. General Dynamics with Pratt and Whitney, worked very closely together to overcome some of these problems which had, in a number of cases, led to the unfortunate loss of aircraft. The flow of air through an engine must have a smooth passage, but it seems that in the case of the F-111, the boundary layer air from the forward part of the fuselage was entering the engine intakes. This could cause the compressor to stall with some rather disruptive results! The engine intake was redesigned considerably, but even today the F-111 still experiences problems. The boundary layer air on TSR-2 was prevented from entering the engine intakes by the use of a large splitter plate. This plate was mounted on a form that was designed to vent the boundary air up and over, as well as down and away from the fuselage.

The Olympus engine has become one of the most versatile gas turbine engines in use today. The various applications in which this engine is used demonstrates its reliability, applications that include power for many of the Royal Navy's frigates and aircraft carriers. Many power stations throughout the world use the industrial version of the Olympus to generate electricity. Perhaps the engine's biggest claim to fame is that it powers Concorde, the World's only supersonic airliner. The decision to choose the Bristol Siddeley Olympus to power the British and French supersonic airliner was indicative of the amount of confidence that had been built up in the engine over a period of time, especially toward the latter stages of its development during the TSR-2 project. This must serve as a tribute to the determination and dedication of the men at Bristol Siddeley, who made it all possible. The Concorde engine, a derivative of the 320 power unit, was designated the Olympus 593 and in cooperation with the French Aero Engine manufacturer, SNECMA, has pioneered supersonic passenger travel during this century.

# CHAPTER FIVE
## The Testing Team

The testing of any new aeroplane is dangerous and an operation where progress is slow, but in speaking to some of the men who take on this formidable task, it is to them 'just a job'. As today's aircraft become more complex, then so is the job of testing and checking out the many pieces of intricate equipment that make up a weapon system. This was certainly the case with the TSR-2 and for this reason there were to be nine aircraft available during the development phase. XR219 was the first and its task was primarily to see if what the designers had worked out, would fly the way they had predicted. Today of course, this task is made safer with the introduction of a new generation of super computers which make it possible to simulate hitherto unknown conditions. This in itself does not take the danger factor away from the man who will be sitting in the driving seat.

Combined with the timetable for the development of the aircraft, was a similar schedule for training those personnel who were to carry out the actual work of testing the aircraft. This group was made up of a project leader, usually the Chief Test pilot, a Deputy, and in the case of TSR-2 with nine aircraft in the programme, would also have involved a number of other company pilots.

When talks began on the conception of the British Aircraft Corporation, Roland Beamont was Senior test pilot and Head of Flight Operations with English Electric. His considerable experience on supersonic flight had made him the ideal candidate to take charge of test flying this new project, under the banner of BAC. In early 1962 Beamont had been nominated by the Ministry of Defence, the Air Ministry and the Board of the BAC companies as the most suitable person for the job. His decision to go ahead and accept the position has now evolved and many of his experiences during those months of testing are covered in various books and articles he has written on the subject.

**The Pilots - Roland Beamont.**
Roland (Bee) Beamont was 18 years of age when he took a short service commission in the Royal Air Force. His flying career began at No 13 Reserve Flying School, White Waltham in 1939, flying DeHavilland Tiger Moths. At the outbreak of war he was completing his training flying Hawker Harts at No 13 Flying Training School, Drem in Scotland. It was shortly afterwards that he took up his first posting as a student pilot at No 11 Group Fighter Pool, St Athan, where he converted to Hawker Hurricanes. In November 1939 Bee was posted abroad to No 87 Sqn on Hurricanes at Lille Seclin in France. Here the squadron formed part of the Advanced Air Striking Force, and he saw action in both France and Belgium, before returning to RAF Hendon after the Allied collapse. Back in England, Bee rejoined 87 Sqn reforming in the North of England and later moving to Exeter to fight throughout the Battle of Britain. In August 1940 he was mentioned in Despatches. During the early part of Spring 1941, whilst operating from Warmwell, Bee took part in the first offensive sweeps over occupied territories and was awarded the DFC. After two years of flying Hurricanes in battle, Bee was to take a short but well earned rest.

As was usual in those days, selected service pilots were seconded to the aircraft manufacturers to assist with the testing of production aircraft. This was to give him his first taste of test flying. Bee was attached to Hawkers at Langley as a Service

# TSR-2 - PHOENIX OR FOLLY?

production test pilot flying Hurricanes and the new Typhoons. After six months he was posted to No 56 Sqn, newly reformed with Typhoons and later on to No 609 Sqn, also with Typhoons, first as a Flight Commander and then as Commanding Officer. It was during his time in command of 609 Sqn that he developed the art of 'train busting' by day and night, himself attacking 25 of the first 100 trains attacked by his squadron in the first three months of their operations. He was awarded the DSO and a Bar to his DFC at the end of his seven months as CO of 609 squadron and he returned to test flying at Langley where he spent another six months testing the new Hawker Tempest. He returned to operational flying in time to form and command the first Tempest wing at Newchurch for the invasion operations and shot down an ME109, the first enemy aircraft to fall to a Tempest, three days after D-Day. Then came the V1 flying bomb attacks on London with Bee destroying 32 of these missiles out of the 630 destroyed by his wing. He was awarded a Bar to his DSO and then in September led the wing forward across Europe to the Dutch airfield at Volkel, from where he shot down his last enemy aircraft, an FW190 on 2nd October. He was shot down and captured during ground-attack operations over Germany on 12th October 1944 and spent the rest of the war as a POW.

After repatriation, Bee was posted to command the Air Fighting Development squadron at the Central Fighter Establishment. In 1946 he left the RAF to become an experimental test pilot with the Gloster Aircraft Company. Roland Beamont had had his first experience of jet flight when he flew a Meteor I of 616 Sqn from Manston in August 1944. Even at such an early stage in its development, and flying an aircraft with qualities that were not particularly endearing, Bee had realised the potential of the jet fighter. Then, in 1947 he became involved in testing the record breaking Meteor IV. In determining the optimum conditions which might be

Roland.P.(Bee) Beamont Chief Project pilot TSR-2 and Head of Flight Operations inspects the cockpit of TSR-2. Along with the Canberra and P1, Bee flew the initial flights in TSR-2 XR219.
*( R.P.B Collection)*

# THE TESTING TEAM

*The Meteor test flown by Bee, shown here on its record breaking run over Bognor Regis. (Rolls-Royce plc)*

encountered during the RAF's official run, Bee achieved 632 mph in this aircraft. Shortly afterwards he moved to De Havilland's as a demonstration pilot.

Later, in preference to accepting an offered permanent commission in the RAF, Bee joined the English Electric Company at Warton in 1947 as Chief Test Pilot. Here he was to lead their new B3/45 (Canberra) jet-bomber test programme. His time spent in 1946/47 test flying the Meteors and Vampires was certainly to help in the ensuing years. An important milestone in his career was reached in May 1948 when Bee was sponsored by the Ministry of Aviation on a visit to America. This was to give him some significant experience on swept-wing aircraft flying in the prototype North American XP-86, (later known as the F-86 Sabre), which he had dived to transonic over Muroc, California. In May 1949 he made the first flight in the Canberra having previously carried out all the ground runs. During these runs Bee had used the long Warton runway to carry out straight hops, reaching a height of about 15ft in order to establish the three axis control and stability of the Canberra. Bee went on to manage all the prototype tests on all major variants of the Canberra and in doing so established two Atlantic speed records in the aircraft, including the first flight by a jet aircraft to cross the Atlantic both ways in the same day.

Perhaps the English Electric P1 was to be the most significant aircraft Bee was to become involved with. Now considered something of an expert on swept-wing aircraft after his single transonic flight in the XP-86, Bee undertook the P1 supersonic flight test programme. On only the P1's third flight he flew the aircraft at Mach 1 on 'cold' engines, (engines without reheat). This was the first British aircraft to achieve supersonic speed in level flight. During this period he became

*The P1 at Boscombe Down. Bee undertook the P1 supersonic test flight programme in this aircraft. On the third flight he flew the P1 at Mach 1 with out the use of reheat. (R.P.B Collection)*

# TSR-2 - PHOENIX OR FOLLY?

the first pilot to fly a British aircraft at twice the speed of sound, the P1-B Lightning. In 1958 Bee was again to visit America specifically to evaluate some of the 'century' series supersonic aircraft. Here he was to fly the F-102 A and B and the notorious F-104, as well as the XF-106. Roland Beamont was the first British, civilian test pilot to fly the F-104 and his recollection serves as a chilling reminder of the danger involved in the work. In preparation for the flight Bee had been given a brief period on an F-104 mockup trainer, followed by an intensive briefing session on the various aspects of the aircraft's handling. Included in his brief was a talk from the spinning project pilot, Dave Hollman on the aircraft's high tail, pitch-up characteristics at the stall. The brief was frank and to the point, with Hollman basically saying 'don't do it'. With this in mind Hollman left to carry out a test flight of his own in an F-104. In the afternoon Bee busied himself getting ready for his flight in the F-104. Firstly he would have a cockpit briefing and then fly. As he walked out to his aircraft a pall of smoke hung over the desert indicating a recent mishap. The accident had claimed the life of the pilot who had just briefed him on the stalling characteristics of the aircraft, Dave Hollman. Bee's first flight was also not without its incident. Having just lifted off from the runway, he noticed a zero pressure reading from the single engine oil pressure gauge. With the engine still sounding healthy he elected to return to base immediately in the hope that this was merely a fault in the gauge. Fortunately he was proved correct, and after this incident he went on to fly two further successful sorties in the F-104 out to Mach 2 over the Mohave Desert.

The P1 had proved beyond doubt that this configuration provided an ideal basis for a new supersonic fighter for the RAF. By 1958 the P1-B had entered flight testing and it was again in the hands of Bee to ensure the programme was as successful as the Canberra had been. Bee was involved in all phases of the Lightning development and indeed on numerous occasions had instigated design changes that enhanced the handling and performance of the aircraft. In early 1957 he had quickly brought to the attention of the designers and engineers the handicap of the aircraft's poor range due to its limited fuel capacity, which ultimately led to the adoption of an enlarged belly tank on the Lightning Mk6 and subsequent variants. He had also wanted a single piece bubble canopy for the Lightning to increase all round visibility, but on this he was overruled, because of lack of confidence in the early 1950s in the strength of Perspex transparencies for aircraft with high supersonic performance and the problems associated with high temperatures at twice the speed of sound.

The formation of the British Aircraft Corporation was to cause Roland Beamont

*The ultimate Lightning. Lightning F Mk6 XR770 of 11 squadron RAF Binbrook. From the P1 through to the F6 Lightning, Bee had done test flying on all marks of this aircraft. (Author)*

to make some important personal decisions. The BAC Board had made a recommendation to the Ministry that Bee should be selected to take on the responsibility of flight testing their important new programme. This indeed was a new challenge which he could hardly refuse. There were however, a number of personal considerations he had to make. It had been his intention to retire from test flying at the age of forty, equally there were other personal circumstances he had to consider. He was given only 24 hours to reach a decision, which at the time, because of circumstances, had to be made alone. Bee therefore had to rely on the understanding of his family when he elected to go ahead and accept the post. This was to be another major achievement in his test flying career: being associated with the TSR-2 programme from its inception, and dealing with all design and development problems until its political cancellation.

In 1965 he became Director Flight Operations (BAC). His cockpit became a desk, and although he continued to do the occasional test flights, in March 1968 Bee was to make his last Lightning test flight after 12 years continuous Lightning testing. This was done in one of the Lightning MK53 aircraft purchased by Saudi Arabia under one of Britain's most successful defence programmes, operation 'Magic Carpet', which today, nearly 30 years later, continues as 'Al Yamamah'. Bee had been one of the founder members of the team which set up the defence programme in Saudi Arabia. In 1968 he was awarded the CBE for his services to test flying. Among other awards for his test flying career he received the OBE in 1953, Britannia Trophy (1953). The Derry Richards medal (1954). The R.P. Alston Medal (1960) and the British Silver Medal for Aeronautics, 'for his contribution to supersonic flight testing' (1965).

1971 saw another change in his distinguished career with him becoming the first Director Flight Operations, Panavia Aircraft GmbH. Here Bee was in charge of Tornado flight testing for British Aerospace and Panavia until the introduction of the Tornado into NATO service. During this period in 1977, Roland Beamont was appointed Deputy Lieutenant for Lancashire.

Since his retirement in 1979, Wing Commander R.P.Beamont CBE DSO* DFC* DL FRAeS RAF (Retd), has taken up aviation authorship and has had published, eight books and many magazine articles. He is a Fellow of the Royal Aeronautical Society and Honorary fellow of the Society of Experimental Test Pilots (USA).

**The Pilots - James Dell.**
There were two other pilots who were to be involved in the initial flying of the TSR-2, Wing Commander Jimmy Dell and Don Knight. You have to describe Jimmy Dell as a quiet unassuming gentleman. His apparently placid exterior manner hides the multitude of stories that portray a career of hard work. He does, however, point out that it involved a certain amount of 'being in the right place at the right time'. This does not hide the fact that without that special flair and abundance of knowledge, along with tremendous hard work, Jimmy Dell would have been just another pilot. The trials and tribulations of his flying career could fill volumes and his escapes from troublesome aircraft alone, have often been highlighted in a number of publications written on test flying.

Jimmy Dell had long decided that the Royal Air Force would provide him with the opportunity to fulfill a life long ambition, and that was to fly. Therefore in 1942 on his 18th birthday, he joined the Royal Air Force. The following year he found himself in Southern Rhodesia, where he was to carry out his basic flying training. On September 20th 1943, he took to the air in a Tiger Moth, the start of what was to

# TSR-2 - PHOENIX OR FOLLY?

*Jimmy Dell with Don Bowen prior to Jimmy's first flight in the TSR-2. Not only did Jimmy test fly the TSR-2 but also flew chase for most of the flights done from Boscombe. All of Don's flights were carried out from Boscombe. When XR219 transferred to Warton Don continued his work at Weybridge. (BAe)*

be a 28 year flying career.

Basic flying training began in Tiger Moths and North American Harvards. At the time the Rhodesian Air Training Group, (RATG), were short of flying instructors and a number were taken from each course and diverted to the RATG Central Flying School to train as instructors. At the end of 1943 with aproximately 250 hours in his log book Jimmy began teaching others to fly. Just after VE Day in 1945 he was back in the UK along with many other surplus aircrew, who had been training in America, Canada and South Africa. After the crew's had attended individual interviews, they were told that because of the huge influx of aircrew from the Empire Aircrew Training Scheme, they were being made redundant from flying. There were however a number of ground posts available, and the RAF would ensure they received all the necessary training. It was flying he really wanted and keen not to miss out on any opportunity, he was one of only three to volunteer out of a bunch of 300 when the RAF stated they still required Elementary Flying instructors in the U.K. This posting was to take him first to Perth and then Rochester. It was here that he saw a signal asking for pilots for a 'Pilot Attack' instructor course on Spitfires and Mosquitoes. Waving caution to the wind and throwing out of the window the old adage, 'never volunteer for anything', he did. This would give him his first opportunity to fly Spitfires and so, before moving to RAF Leconfield where the course was to be held, he had to attend RAF Spitalgate to be checked out on the Spitfire, which he had never flown operationally. At the time the British Government had just sold a batch of Mk 9 Spitfires to the Turkish Air Force and required instructors to train the Turkish pilots. Despite having only just been checked out on the Spitfire himself, he nevertheless, did have an instructor's rating and so instead of moving to Leconfield to do the attack course, he found himself training Turkish pilots to fly Spitfires. He did eventually get to Leconfield and successfully completed the Pilots Attack course and in fact stayed on as an

# THE TESTING TEAM

instructor for a further 3 years until 1949.

His latest posting was to 43 squadron at Tangmere where they had just been equipped with the latest Mk 8 Meteor. Jimmy had already shown that he certainly had potential, and after only 6 months at Tangmere, was sent to West Raynham to attend the Day Fighter Leaders course, where he would be taught how to lead a squadron. Two months after completing the course he was posted back to Raynham as an instructor, a job he was to do up until 1952. Around this time he was selected for an exchange posting with the United Sates Air Force and sent to an F86-E unit in the United States. This was to give Jimmy the ideal opportunity to see the latest technology in the world of military aviation, for halfway through the exchange, the unit he was attached to received the latest F86-D, a single seat, all-weather jet fighter. It was during this particular visit he had the unfortunate experience of having to eject from a troubled F-86E.

This was June 1953 and there was little, if anything, in the Western world that could equal the F-86. Some of its equipment included electronic fuel control, autopilot with auto ILS, zero reader, VOR, pilot operated radar and 2.75 inch rocket armament. The knowledge gained from this visit really established Jimmy's reputation as the only UK pilot with experience of a pilot operated radar system, which, back in 1954 was most significant. Back in the UK Jimmy went to the Fighter Weapons School at RAF Leconfield as a Squadron Leader. It was around this period that the P1 was beginning to fly in the hands of Roland Beamont, initially as an experimental research aircraft and later in direct preparation for its planned successor the single-seat, Lightning.

Contrary to the normal practice of keeping the customer away from the constructor, Jimmy was seconded to English Electric at Warton in 1957 as the RAF Project Pilot for the new Lightning. This was a new idea, completely breaking with tradition, as normally this work would have been handled exclusively by the pilots from Boscombe Down. He reported directly to the C-in-C of RAF Fighter Command and at English Electric to Bee. The concept was to have an operational pilot, (not a test pilot) to join the flight development team as the Operational Requirements Liason Officer, (ORLD), whose job it was to ensure fully the developed aircraft met the services needs. During this three and a half year period he had been promoted to Wing Commander and was posted back to the RAF to take over the Air Fighting Development Squadron at West Raynham, flying tactical trials with the new P1-B Lightning. Prior to this, he was given the task of leading a team to the United States to fly all their supersonic aircraft. This again was an annual visit usually led and carried out by Boscombe Down pilots and yet, here was a man, now a Wing Commander, who admits to never having been on a test pilot's course, leading a team of high ranking, experienced Air Force and Naval officers to fly these complex aircraft. During their stay they were stationed at the Edwards Air Force base and flew the F-100, F-104, F-105 and the F-106, all within a 5 week period. The Boscombe Down pilots were there to evaluate the aircraft and to check performance figures against those of the manufacturers and Jimmy was there primarily to assess the weapons systems.

Just over halfway through this evaluation period, Jimmy received an urgent signal recalling him home. Mystified as to what was happening, he returned only to be told that he was immediately to lead another team from the Central Fighter Establishment, (C.F.E) on an evaluation exercise, this time to Mont de Marsan, one of the French Flight Test Centres. During his time here he was to fly the Vautour 2A, Super Mystere and Mirage 3. It was an extremely busy period, for in the space of approximately six weeks he had flown all these aircraft and was now required to

## TSR-2 - PHOENIX OR FOLLY?

*The F-104C Starfighter in simulated combat mission over the Majove Desert. It was this type of aircraft that Roland Beamont and Jimmy Dell flew during their visits to the U.S.A. (Note the In-Flight refuelling probe on 60908)*
*(Lockheed Corp)*

write a detailed report on each of them. A daunting task that he admits never having completed to this day......

At the end of 1959 Jimmy was back at West Raynham and the job of introducing the Lightning into the RAF continued, which also necessitated a return to English Electric at Warton. In the meantime English Electric themselves had run into problems of their own. Roland Beamont, the Chief Test Pilot was having a medical problem with his neck and DeVilliers, the Deputy Chief Test Pilot, had just had a Gall Bladder operation that had gone wrong. Johnny Squier, who was the Chief Production test pilot, had been asked by Bee to do some of the development test flying on the Lightning. On October 1st 1959, whilst doing a high speed test run up the Irish Sea in T-4 Lightning XL628, Squier got into severe difficulties. Having just completed a high speed roll, the plane suddenly went out of control due to a structural failure, leaving Squier with no option but to eject. Having spent some 28 hours in the sea, he was rescued and confined to hospital. It was eventually found that the fin had broken away. This was a situation Jimmy was to find himself in 1965, whilst test flying the T-5 version of the Lightning.

These circumstances put English Electric in quite a quandary and forced the company to approach the only person whom they felt was suitably qualified to take on the important role of experimental test pilot. That person was Jimmy Dell. Roland Beamont's decision to approach him was based on the quality of his previous work at Warton and his knowledge and experience already gained from his flying duties with the RAF and USAF. However, Jimmy himself was somewhat loathe to leave the RAF as he believed he had one of the best jobs in the Air Force at

*XM968 the Lightning flown by Jimmy Dell for the chase flights. The aircraft was written off in 1977 after a ground collision.*
*(Stewart Scott)*

122

the time, although he did realise there were other things to consider. One of these was that things cannot last forever. He had already done nearly seventeen and a half years as a pilot and had not completed a ground tour, and so he accepted that to some extent the writing had probably been on the wall for some time. One of his close friends was Group Captain Crowley-Milling, (now Air Marshal retd), who was in operations at the Central Fighter Establishment, West Raynham, and it was to him that he went to seek advise on this particular subject. 'Wouldn't touch it with a barge pole old chap', was the advice he received there. Jimmy's Air Fighting Development Squadron was based at RAF Coltishall so he then asked the Station Commander there his good friend, Group Captain Bird-Wilson, (now Air Vice Marshal retd), what he would do. 'Wish they'd offer me the job' was the prompt reply. So, in the space of one day he had received these two wildly conflicting opinions. The decision was entirely in Jimmy's hands of course, and in the end he chose to join English Electric. He took up the position initially as Deputy Chief Test Pilot in December 1959, but just over a year later, in January 1961, when Roland Beamont became Manager Flight Operations continuing as Head of Test Flying, he became English Electric's Chief Test Pilot.

The work he became involved with at English Electric, which at the time was to be mainly on the Lightning, would provide obvious advantages for the Air Force as he understood the priorities and needs of the RAF. It was this combination of interests that worked extremely well for both the RAF and English Electric.

His involvement with the development of the Lightning brought him not only a great deal of satisfaction but also a few narrow escapes. During English Electric's great Saudi Arabian defence programme, Airwork were to assist with some of the deliveries. This required some Airwork pilots to be checked out on the type. On March 7th 1967, Jimmy flew a familarisation flight with Peter Williams in a T-55, (the Saudi version of the RAF T-5). The flight was from Warton, up into the Scottish Highlands, and then back down over the Pennines, returning to Warton. When they took off, a strong wind was blowing across the runway but was well within the crosswind limits set down for the Lightning. On their return to Warton they were advised by Air Traffic Control that the crosswind had increased to well over 30knots. This was now well outside the crosswind landing limits for the T-55, but Jimmy had insufficient fuel to divert to another field. By now time and fuel were running out and he had no option but to put the aircraft down. As soon as the wheels of the aircraft touched the ground, Jimmy attempted to bring it to a halt as quickly as possible and he managed to keep it on the runway for approximately 500 to 600 yards. Then the wind began to push the aircraft toward the soft ground at the side of the runway. He applied the brakes harder, leaving skid marks as evidence of his attempts at keeping the aircraft on the runway. Unfortunately the Starboard wheels ran off the edge of the runway into the soft earth and began to sink in, pulling the Lightning round even further. The nosewheel struck a huge concrete base at the side of the runway, snapping the nosewheel off and flinging the aircraft into the air. The Lightning then crashed down to the ground, with the nose intake digging into the soft ground. The momentum slewed the whole airframe round severing the cockpit from the fuselage at the cockpit bulkhead, which had already been weakened by the smash into the concrete base. When the dust settled, Jimmy was still in his seat but his left leg was badly twisted and in the area where pilot's seat should have been. The emergency services arrived quickly and doused the damaged aircraft in extinguishant. Peter Williams was nowhere to be seen and at first it was thought he had somehow ejected, but when the crash crew began looking round the damaged plane they suddenly discovered him, still in his seat and alive,

## TSR-2 - PHOENIX OR FOLLY?

*The incident when Jimmy was forced off the runway at Warton by a strong crosswind. The grooves in the soft earth clearly show the track of the T55. Jimmy and his passenger, Peter Williams were lucky to escape with relatively minor injuries. (BAe)*

under the port wing. Peter Williams had suffered a broken thigh, and Jimmy's left knee joint was smashed.

During this period in the development of the Lightning, the TSR-2 came along and although Jimmy was now the company chief test pilot, from the begining of the programme the English Electric Board had directed that Roland Beamont would undertake the initial critical flight testing on the TSR-2 at Boscombe Down, and would, when he considered it fit, bring XR219 back to Warton. Then when an appropriate standard of safety had been achieved Bee would hand over the TSR-2 test flying programme to Jimmy. He had already given his full support to this arrangement. This was to be Bee's final project as a world renowned test pilot and, Jimmy would at the appropriate time be ready to carry on the Warton test flying tradition in his own right.

His experiences whilst testing the TSR-2 are covered later on, but after the cancellation of the whole programme attention was quickly turned to the Anglo-French project that so far had been at the discussion stage, the Sepecat Jaguar. The Jaguar was a completely different aircraft from the Lightning and the TSR-2, though BAC did find its experience gained from the TSR-2 useful.

Because the in-service date of the French Jaguar was to be earlier than that of the Royal Air Force, it was agreed that Jimmy would join in test flying the Sepecat/Breguet prototypes at Istres, the French Flight Test Centre, on the Southern Coast of France. There were to be five prototypes and Jimmy would carry out the BAC basic handling trials. By the time these trials had started it was hoped the British version, with its more complicated avionics, would be ready. He would then return to the UK and continue the development flying of the Jaguar at Warton. By now of course the Multi-Role-Combat-Aircraft, or MRCA, later to be known as the Tornado, was beginning to come together and Jimmy started to become heavily involved with this project also.

On a Tuesday in December 1970, Jimmy had been for his usual six monthly medical. This included an Electro-Cardiograph and was carried out by the company doctor. The immediate results had shown no defects and therefore the doctor had signed him as fit to fly fast jets. Further detailed analysis of his ECG would be done by other specialists. The following Saturday, Jimmy was once again in the seat of a Jaguar doing a number of test flights. They were, he recalls two of the most enjoyable and satisfying flights he had ever done. On landing back at Warton he

## THE TESTING TEAM

was told there was an urgent phone call for him in Flight Ops. It was the company doctor, who immediately ordered him not to do any more flying and that he would be over to see him straight away. When the doctor arrived he told Jimmy that the specialists who had studied his medical results had found an irregularity in his ECG. A normal heartbeat, comprises of a primary beat followed by a softer secondary beat, but it was found on Jimmy's ECG scan that one of the secondary beats was inverted and was below the normal line. Even though he had a number of medical examinations and consulted various specialists, including the Senior RAF medical officer, he was told that his days of test flying were over.

This was a blow to Jimmy and his immediate reaction was to turn his back on aviation totally. He felt that if he could not fly, then he did not want to be around aircraft. However, BAC needed men of his calibre to help with the new projects that were up and coming. His considerable knowledge and experience could be put to good use in the problem areas that lay ahead. Roland Beamont, now Director of Flight Operations and the International flight test management team, recommended him for role of Manager Flight Operations for the new Tornado programme, and Jimmy said that he would do it, but only for a short period until such time the company could find a replacement. This period was not so short and was to last nigh on 15 years, firstly as Manager Flight Ops for 5 years, and then after Bee's retirement in 1979, a further 10 years as Director Flight Operations.

During this time he had a number of medicals and each time was passed fit enough to fly and in fact flew the Tornado on a number of occasions. Although he had not flown a fast jet for some time, he found he could still fly the aircraft with reasonable accuracy. All the other important duties a pilot has to perform whilst flying, such as talking to the ground control, changing radio frequencies and handling the radar, had gone completely to pot, (he was well out of practice). But it was good to get back into the driving seat again and he enjoyed it tremendously. Jimmy has spent all his working life in aviation enjoying virtually every moment, and even now he still often finds it hard to believe why they actually paid him to do a job he so thoroughly enjoyed. Today Jimmy Dell OBE, has many momento's of those years spent in the flying seat. He has been awarded the Derry & Richards

*A Sepecat/BAC Jaguar takes off from RAF Coltishall. The Jaguar benefitted from the building of the TSR-2 in that many lessons had been learnt. The Anglo-French programme had also gained from the Concorde project. (Rolls-Royce plc)*

# TSR-2 - PHOENIX OR FOLLY?

XR219, Jimmy Dell, Don Knight, Roland Beamont and Don Bowen. Missing are Peter Moneypenny and Brian McCann, both of whom were busy that day. This shot was taken shortly after the cancellation. *(BAe)*

Award, (Guild of Air Pilots & Navigators) and the Royal Aero Club Silver Medal for his services to test flying. Today, even though he is retired, his help and advice is often sought by companies and organisations throughout the land.

**The Pilots - Don Knight.**
The third pilot to fly the TSR-2 was Don Knight, a tall distinguished Scotsman. He first started flying in 1949 as a National Service pilot in the RAF. Don's basic flying training began on Course Number 34 at RAF Feltwell, flying the Percival Prentice. After some 100 hours on the Prentice, Don moved on to the North American Harvard 2-B for the applied phase. From the Harvard he then went on to advanced training, flying Meteor 4's and 7's whilst stationed at Middleton-St-George. His Operational Conversion came at RAF Stradishall, again flying the Meteor. On completion of Operational Conversion, Don, now a Flying Officer, had duly completed his National Service and in April 1951 left the RAF to return to college in Edinburgh. Here he was to study printing technology, a complete change from the work he had been doing whilst in the RAF. He did not completely sever his ties with flying, mainly because of his reserve commitment and joined the Royal Auxiliary Air Force, number 603 squadron, City of Edinburgh. Almost immediately he was called up to attend a 3 months refresher training. Apart from this Don flew most weekends and during the summer camps, flying initially in the Spitfire 22 and then converting on to the DeHavilland Vampire 5.

During these weekends spent flying Don found new friends, and soon learned that a number of his colleagues were seeking positions in the aviation industry. It was early in 1953 when Don himself heard that English Electric were expanding and that they had a vacancy for a Production Test Pilot. At the time Don had no experience as a test pilot, not having attended any of the recognised establishments. He had a good service flying record and he had won a number of cups as a student pilot, both for flying assessment and ground school subjects at Flying Training

# THE TESTING TEAM

School. This was good enough for the man who was eventually to employ him. Bee's policy when employing test pilots at that time was based on the man's flying assessment and personal qualities rather than the tradition of employing only pilots who had attended the RAF's Empire Test Pilots School.

So in June 1953, at the age of 22 years, Don Knight joined English Electric as a part-time production test pilot working under Johnny Squier. He was also appointed to fly the company's communication aeroplane, at that time a DeHavilland Dove. Not long afterwards he started full time testing as the company continued to expand. A fellow colleague from 603 squadron, Tim Ferguson, was soon to join English Electric and took over as the part-time communications pilot. This left Don free in 1956 to move from Samlesbury to Warton to take up full-time work as part of the Warton team of experimental pilots. The work being carried out at the time involved Don mainly on Mark 8 Canberras in development and autopilot clearance for the Indian Air Force. Other work included the Mark 9 Canberra aerodynamic and systems development.

The P1 and P1-B programmes were some three years old when Don became involved in the Lightning project. Up until this time he had no experience on swept wing flying and was duly sent off to the Rolls-Royce Flight Test Establishment at Hucknall in Nottinghamshire. This was the period when Rolls-Royce were involved in the intensive flying phase of the RA28 Avon programme using the Hawker Hunter Mk-6. Don put in a number of hours in this plane checking out the engine and its handling. During some of the flights Don had to shut the engine down and found himself over RAF Waddington simulating engine failure in order to assess the engines startup characteristics. The Hunter of course is a single engined aircraft and at first Don found this a little unnerving! There followed an increasing involvement in all aspects of the Lightning aerodynamic systems and weapons clearance programmes, and much to do with the new generation OR946 flight control and instrumentation system, and the Lightning autopilot clearance.

Don's first involvement on the TSR-2 project came about when a concensus of pilot's views and opinions were sought on various aspects such as the cockpit layout. For the first six years of the programme Roland Beamont, with Jimmy Dell as his deputy, had the flight operations responsibility for the TSR-2 project. However, when Jimmy took on the responsibility of flight testing the TSR-2, in the last two months before cancellation, Don became more involved, basically as Jimmy's No. 2. Before that and before first flight, Don studied aspects of the flight simulation programme which culminated in him being selected as the evaluation pilot on a research project to Cornell University at Buffalo, New York in 1963.

The university had a T-33 which was considerably modified and converted into a variable stability research aircraft. As such the aircraft was able to simulate a wide range of aerodynamic configurations. This was a tandem two seater aircraft with the controls from the front cockpit going through an analogue computer and a set of servos, which would allow the pilot in the back to select certain parameters that would simulate specific aerodynamic characteristics. The rear cockpit had controls routed normally to the flying surfaces for safety reasons and to ensure control should the man in the front get into trouble.

This was to give Don an insight into the configuration of the TSR-2 under certain conditions. Although the programme was classed as 'general research' for political reasons, the case studies included examples of the expected lateral stability parameters of the TSR-2 in the landing configuration. In all, 67 different aerodynamic configurations were flown. It was very shortly after this that Don became involved in the multi-axis flight and systems simulator which was then

# TSR-2 - PHOENIX OR FOLLY?

*The simulator at Weybridge an important part of the project that Don Knight became involved in. (BAe)*

being constructed at Weybridge. Another of his involvements was with the terrain following radar, which was to be fitted in XR221 but was not yet ready for actual flight. As soon as this aircraft joined the development programme, Don would have become much more involved in the development flying of the TSR-2.

Don first flew a familiarisation flight (Flight 11) in XR219 from Boscombe Down on February 10th 1965. The details of his flight are covered later on. His impression was, not surprisingly, similar to that of those who had previously flown the aircraft. One of his comments concerned the view from the cockpit. Normally the pilot could see the aircraft's wings, and was also able to see some part of the nose. This allowed him to use it as an attitude reference. The long fuselage on the TSR-2 meant the pilot, when sitting in the normal position, was unable to even see the wing tips, and the nose also sloped away so sharply that he would be unable to use it for reference. This would pose no problem to pilots used to flying the Lightning as the seating position was very similar.

After the project was cancelled Don stayed with BAC and became involved with the early part of the Saudi Arabia defence programme. He delivered the first Lightning to Riyadh. He set off with Jimmy Dell leading in another Lightning and they flew, firstly to Akrotiri in Cyprus in-flight refuelling on the way. After an overnight stop in Akrotiri the pair flew on to Riyadh. Jimmy the leader, was to land at Riyadh first, but due to a problem with an in-flight refuel he had to divert to Amman and Don carried on to Riyadh. After this he delivered two other Lightnings out of the initial batch of six.

Don continued testing the Lightning and played a significant part in the stalling and spinning programme which was led by Jimmy Dell. Then the Jaguar project started and again Don became involved, mainly with the cockpit layout but he also took part in talks with Rolls-Royce with regard to the engine. In 1967 he had to retire from active test flying due to medical problems, but stayed with BAe at Warton in the Sales and Marketing Division as BAe's Marketing Director, selling Canberras, Strikemasters and later still, Jaguars. In addition to the many other awards from his

# THE TESTING TEAM

early days of flying, Don was awarded the Queens Commendation for services to flying.

**The Navigators - Donald Bowen.**
The man whose job is done from the back seat, the navigator/observer, tends to be overshadowed by the more glamorous job of the pilot. Don Bowen was the Chief Project Navigator (TSR-2) who flew with Roland Beamont during the early, critical, testing phase of TSR-2. When XR219 moved to Warton, Don stayed at Weybridge working on the navigation equipment that was destined for the aircraft. This was important work and Don made an essential contribution particularly, in the early testing phase to the TSR-2 programme.

Don had always taken a keen interest in aviation and even at the age of 8 had his first flying experience in an Avro 504K from Croydon. During the second World War in 1940, Don enlisted in the Royal Air Force for flying duties. He went on to train in South Africa, under the Empire Training Scheme, and soon Don became a qualified navigator, subsequently serving with Coastal Command where he flew navigator in B-17s, Hudsons and Liberators. It was soon apparent that Don Bowen had natural ability as a navigator and after attending a specialist navigation course at Shawbury, he went on to become station navigation officer in RAF Transport Command's development unit at RAF Brize Norton.

After the war years Don joined EMI in their guided weapons division. At the time the company were involved in developing the active homing device for the Red Dean, air-to-air missile. The flight trials were carried out using a Canberra, operated by the Vickers Armstrong flight development department from the company airfield at Wisely. Don was eventually to join Vickers and soon became involved in test flying. This also gave him the opportunity to operate delivery flights in Vanguard and Viscount airliners to operators throughout the world.

It was in 1963 that Don Bowen first became involved in the TSR-2 project, joining the programme as Chief Project Navigator. This was also the first occasion that he made the 'first' flight in a prototype aircraft. As an experienced navigator one of his tasks was establishing how the test crew could use the low-level routes and bombing ranges for testing TSR-2 in the United Kingdom. During this period Don spent some time in the USA in order to gain experience with the American terrain following and nav/attack system. The first development TSR-2 prototype would lack this type of sophisticated navigation equipment as well as the auto fuel balancing mechanism. Instead this essential work would have to be carried out by the 'man in the back seat'. The first ground testing was carried out by Roland Beamont and Don Bowen as were the first few critical flights. During flight 5, the undercarriage had failed to lower properly. Bee gave Don the opportunity to evacuate the aircraft before Bee attempted a landing. Despite knowing full well the possible consequences of the undercarriage failing, Don bravely elected stay with the plane. When XR219 was transferred to Warton Don continued his work at Weybridge preparing the third aircraft, XR221. This would carry many of the systems on which he had assisted during development. Only the day before cancellation Don had successfully run a complete sortie on the system.

The cancellation was a great blow to Don and it was to Canada he turned joining Canadair in Montreal. Don began working on the development and flight testing of the first water bomber. His aviation career was completed whilst working with deHavillands as a quality engineering manager. Those that knew him and worked with him acknowledged his loyalty and dedication to his work. Roland Beamont

# TSR-2 - PHOENIX OR FOLLY?

*Peter Moneypenny and Jimmy Dell. A strong friendship between these two gentlemen was formed when Jimmy joined English Electric in December 1959. When Peter suffered a stroke in 1981, Jimmy, although stationed with Panavia in Germany, flew back to England regularly to help his good friend. (BAe)*

praised his professionalism as a member of the TSR-2 flight test crew. He was a tremendous asset to the TSR-2 programme and made a significant and practical contribution to the project.

**The Navigators - Peter Moneypenny.**
There were two other navigator observers, Brian McCann and Peter Moneypenny. Peter Moneypenny, as Chief Navigator (Warton), flew as navigator/observer from both Boscombe Down and Warton, and made an important contribution to the testing of the TSR-2. Peter has had a varied career in aviation, starting way back in 1942 when he joined the RAF for aircrew training having had a brief period in the Royal Navy. He went to Canada to begin basic navigational training and on completion of this he was commissioned and returned to Britain to begin operational training. Here he flew operationally with 115 squadron in Lancasters. The years just after the war were spent in Aden, where he flew on a communications flight and spent time as Command Navigation Officer, HQBF Middle East, also OCRAF Hargesia, British Somaliland. In 1948 he returned to the United Kingdom where he joined the Telecommunications Flying Unit, TFU, at RAF Defford. It was here he was to gain experience flying in a number of different aircraft including the early Marks of Canberra. Peter was also to gain a great deal of experience doing research with radio and radar flying aids, which included firsthand experience on Doppler navigation aids and H2S derivatives. The H2S was a similar piece of equipment to the Matra MARTEL missile, which is covered later.

In 1951 he left the RAF and joined a RAFVR squadron as a navigation instructor, before going on to join Silver City Airways. Here he started a jet ferry unit with John Hackett, delivering Canberras for English Electric to the many air forces throughout the world which had purchased these aircraft. It was during this period when he was to meet someone who was to become a very close friend, Jimmy Dell. Later through his contacts at Warton, he joined English Electric Flight Operations Department and for a while, continued delivering Canberras to Venezuela, Equador, Peru, Ethiopia and India. He also flew on delivery flights with

# THE TESTING TEAM

Strikemasters and Lightnings to Saudi Arabia. A number of these deliveries were done with Jimmy who, particularly with the Lightning and Strikemaster deliveries to Saudi Arabia, would often refer to Peter as his 'automatic pilot', because Peter, having gained his pilots licence, used to do the flying whilst Jimmy sat back and enjoyed the view.

It was during his Atlantic flights in Canberras that Peter, flying with John Hackett, broke a number of Atlantic records, but because they were not flown under full F.A.I rules they were never officially recognised. However, on August 25th 1955, under strict F.A.I rules, John Hackett, pilot, and Peter Moneypenny, navigator, flew from Heathrow Airport overflying Croydon to New York and back again in a day, gaining three records: London - New York, New York - London and London - New York - London. For this they were jointly awarded the Britannia Trophy for that year. It is believed these records still stand today. It was after this that the really serious work began, when Peter first of all obtained his Air Transport Pilots Licence that would allow him to fly both as a pilot and as a navigator.

As Chief Navigator at BAC Warton Peter Moneypenny, like Jimmy Dell, was to become deeply involved in the TSR-2 project at a very early stage, especially in areas of equipment and the layout of the rear cockpit. Although much of the equipment had already been selected, it required experts like Peter and Don Bowen to coordinate the best position for it so that when TSR-2 entered service the navigators could do their job efficiently and comfortably.

Having flown as test navigator on Canberras and Lightnings, it was a natural progression for Peter to become involved in this new project. He first started getting to know the TSR-2 from the flight simulator which had just been built at Weybridge and although it did not look too much like the real thing, was acclaimed by those who used it as a superb instrument in which to gain experience before moving on to the real thing. The simulator was to provide him with an ideal scenario, flying many sorties and training for a role that was never to come. His job was to assess and test the efficiency of the navigation and attack equipment for the man who would have eventually flown in the back seat. The pilot and navigator in such aircraft as TSR-2, work as a very close team. Without one the other cannot fulfill his job properly. The test flying in the prototype TSR-2, XR219, required the navigator to take on the role of Flight Test Observer, with the pilot navigating in the absence of the sophisticated nav/attack equipment, which was to come later. Then as this equipment began to be fitted in the airframes, the navigator's job would become more demanding and exacting. Even so it was never just a case of sitting there watching the world go by. The work load of a Flight Test Observer was extremely high, and none of the three navigators had much time to enjoy the view. During most of the test flights, Peter was mainly assisting with pre-flight checks, in-flight checks, some navigation and of course the all important cross checking of the fuel. One thing Peter Moneypenny can lay claim to, is that he is the only navigator to have done a 'barrel-roll' in the back seat of a TSR-2, for on flight 16 test pilot Roland Beamont, had been carrying out Dutch rolls and partial rolls through 90° and 180° on XR219, and then at around 5,000ft carried on to roll through 360°. This was performed after checking instrumentation with Peter in the back seat.

Today Peter Moneypenny lives in a quiet village in Devon, having suffered a stroke in 1981. One of the first people to come to Peter's aid was his good friend Jimmy Dell. The tremendous rapport between these two is amply demonstrated by Jimmy's insistence on visiting Peter even though he was based out in Germany with Panavia.

# TSR-2 - PHOENIX OR FOLLY?

## The Navigators - Brian McCann.

Brian McCann considers himself the junior partner in the team that flew XR219, with his introduction to the TSR-2 project coming towards the end of the flying programme. He joined the RAF at the age of 18 on the 1st December 1947 on an 8 year Short Service Commission. He went through Grading School at North Weald, passing the aptitude and written tests with flying colours and was soon able to realise his ambition of becoming a navigator. Initial training was done in Southern Rhodesia on Avro Ansons before returning to England in order to complete his training on Wellingtons and Mosquitoes. Training completed, Brian was posted to 39 Sqn flying Mosquito night fighters based in the Suez Canal Zone. In 1953 Brian once again returned to the U.K to join 25 Sqn at West Malling where he was to fly in the twin seat Vampire NF-10 and Meteor NF-14. After attending the Staff Navigation Course at RAF Shawbury, Brian completed his service as the Station Navigation Officer at the Night Fighter O.C.U, RAF Colerne.

As with most men whose careers have been spent flying around the world in aircraft belonging to Her Majesty's Royal Air Force, Brian sought a civilian job that would allow him to do similar type of work and also give him that same kind of excitement. His first venture into civilian flying was with AVRO at Woodford in Cheshire, as a Flight Test Observer. Here he joined the team developing one of Britain's stand-off bombs, which was eventually known as Blue Steel. The work was to be a little disappointing, mainly because he found that the company were employing flying staff well in advance of the flying programme. The bomb itself was not due to be flown for at least another 2 to 3 years and Brian was in the second intake of crew awaiting their training. So, after only 6 months, Brian decided he wanted something a little more exciting and duly applied for the position of company navigator with DeHavilland at Hatfield to replace the RAF project navigator.

The work at Hatfield was to involve a new air-to-air missile and this gave Brian the opportunity of flying in the many different types of aircraft that were used in the development programme. These included the Comet 1, Canberra, Venom and Meteor. The work was a little more demanding than that at AVRO and Brian recalls many an exciting time spent over the Aberporth range firing the new missile. He remembers one particular incident when, flying in a Venom, the pilot fired the missile only to have it break up immediately in front of the aircraft. Fortunately they flew through the debris without any of it being ingested into the engine. Brian stayed with DeHavilland on the missile development programme until its fruition, when it became known as the Red Top intended for the RAF's Lightning.

After his time at Hatfield Brian began looking for a job that would give him an opportunity to work on test and development aircraft and this eventually led him to English Electric at Warton. This was to be the start of a 15 year attachment with the company. It was around this period that the Lightning T-4 was being readied for its first flight and Brian was eager to take part in the development of this new aircraft. However, it was not to be for some time as the T-4 was essentially a two pilot

Brian McCann was the third navigator/observer in the TSR-2 programme. Pictured here prior to his first flight in TSR-2. Later Brian changed seats and became a flying instructor then a senior airline captain. (B.McCann)

132

aeroplane. Brian's time with the Lancashire company was spent mainly in flying in the Canberra, both production models and often types testing new equipment. Other work was to involve him flying in the Jet Provost and later its export version, known as the Strikemaster. The two seat Jaguars and, perhaps Brian's favourite, the T-5 Lightning were other aircraft with which he became involved whilst working at Samlesbury and Warton as well as with the more mundane task of navigating the company's communication aircraft, the Dove, Heron and HS-125.

When the TSR-2 came along Brian perhaps knew he would eventually become involved, but it was not until he was in Venezuela, having flown as navigator delivering a new Canberra that he was actually invited to join the TSR-2 team as a navigator/test observer. He was waiting to bring another Canberra back to Warton when Peter Moneypenny flew in navigating another Canberra on delivery. Peter at the time was the Chief Navigator at Warton and basically Brian's boss. The two men spent a few days together before Brian was due to fly back to Britain, and Peter asked Brian if he would like to join the TSR-2 team. This was sometime in 1963 and well before the first flight of the TSR-2. Brian knew that this was the company's main military project at the time and of course accepted straight away.

Brian's involvement in the initial stages of the development of the TSR-2 was minimal. The nav/attack systems were to be developed at Weybridge and there was little indication then that this work would be moved to Warton or Samlesbury. He spent most of his time learning what he could about the project, which at the time was very difficult as most of those already involved were still busy establishing themselves on the new project. He did however sit in on all the pre-flight briefs and de-briefs, learning and understanding the duties of the navigator, whilst still being heavily involved with other company work. For Brian March 1965 was an extremely busy period in his flight testing career, for not only was this the month he did all his flights in the TSR-2, but he also was carrying out E.C.M, (Electronic Counter Measures) trials in Canberras at the military ranges over the North Sea.

XR219 had very little in the way of navigation equipment, most of the space in the navigator's cockpit was taken up with engine instrumentation and equipment to measure and study the fuel flow. Brian's first flight in the TSR-2 XR219, was on March 8th for flight 17 with Jimmy Dell as pilot. For Brian it was an introduction to TSR-2, and he remembers very well coming back with reams of notes. All of Brian's flights in XR219 were with Jimmy as pilot.

Not all of Brian's work on the TSR-2 was in the aircraft itself, and neither was the job of Flight Test Navigator always done in the comfort of a nice warm cockpit. One of the trials to be carried out on one of the aircraft, XR223, was flying the aircraft with both canopies removed. This was to assess the effects not only on the aircraft, but also on the crew after canopy ejection. Before any aircraft was flown without its canopy it was essential to simulate these conditions on the ground in order to discover if such a test would endanger the crew or the aircraft. To carry out this type of simulation Boscombe Down had a huge instrument of torture. It was a large fan powered by Rolls-Royce Merlin engines that would blow a slipstream of air at speeds to over 400 knots over a mock-up of the TSR-2 cockpit, minus the canopies. Brian remembers this nauseating experience as he sat in the back cockpit with a 'chicken out' button he could press that would stop the fan, should he find the experience unbearable. The worst effects, he remembers, were at the high speed end which had a disorientating effect in very heavy buffeting. Beamont, Dell, Moneypenny and other Warton aircrew had also been subject to this particular form of 'torture'.

After the cancellation of the TSR-2 programme Brian stayed on at Warton to

# TSR-2 - PHOENIX OR FOLLY?

work on the Jaguar and eventually on preparatory work for the Tornado. In 1972 Brian, with Paul Millett, brought the second development French Jaguar E02, from Istres in France to Warton, a distance of some 600 miles. His work for Tornado included its low level route assessment, something he might one day have been doing on the TSR-2.

Brian McCann credits the 'boss', Roland Beamont for the way he held his flying team together after the TSR-2 cancellation. Many hundreds in BAC at the time were being made redundant, but it was Bee who insisted on holding on to his flying staff until work that had previously been contracted out to other companies could be brought back to Warton.

After completing almost 20 years in the aviation industry as a navigator Brian 'changed seats' and went on to become a flying instructor followed by many years as an executive pilot. In 1990 Brian finally hung up his goggles, having risen to Senior Captain with an executive charter company based at Heathrow, a far cry from those exciting days of 1965. What was Brian's impression of the TSR-2? It was he says, *'a great aeroplane, a threat not only to any potential enemy, but also a threat to the American aircraft industry'*.

## Ground Crew - Len Dean.

The man in charge of the TSR-2 ground staff, particularly during the period it was at Boscombe Down, was BAC's Chief Ground Engineer, Len Dean. Len was responsible for ensuring all the systems and components, including the engines, were checked out fully prior to the first flight. At this phase of the programme it was

Looking very pleased Len Dean (in white overalls) explains some part of the TSR-2 to invited guests at Boscombe. Len had done similar work with Bee firstly on the Canberra at Warton and then on the P1 at Boscombe. *(BAe)*

essential that everything went with as few hitches as possible and therefore it was part of his job to ensure a smooth transition. Len co-ordinated the engine ground runs with the Bristol Siddeley engineers, and his own men from BAC.

Len Dean is a very proud and conscientious man. He has a lot to be proud of and was always aware of the importance of his work. Len started his working life at the age of 14, when he left school and joined English Electric as an apprentice fitter. Three years later Len worked as a flight engineer on Halifax MK2, 3, 5 and 6's between 1940 and 1945 after the company had won a contract to manufacture the aircraft. He then went back to engineering first by working on the DeHavilland Vampire and eventually in 1949, becoming involved with the prototype Canberra working alongside Roland Beamont. This was not the first time he and Bee had worked together. Some years before, whilst Bee was working at DeHavilland, Len had met him. In 1947 Bee had joined English Electric to take charge of the Canberra test programme, and it was from then that Len began a long association working with Bee. After the Canberra came the P1 and Len recalls the days spent at Boscombe Down again with Bee getting the P1 ready for its first flight.

When the TSR-2 programme began, Len took charge of preparing this new and important aircraft for its first flight. Initially he had busied himself finding out what he could about the aircraft and its intricate new systems, new engines, new electronics and new airframe. Because of the known problems with the engine, he felt it essential for an efficient means of escape to be established for the men who would be sitting in the aircraft during the engine ground runs, in the very likely event of a catastrophic failure. He also wanted the engine to be physically checked and sent an engineer down the engine intakes to check the compressor blades before it was run.

Although Bristol Siddeley were responsible for the engines, Len would still ensure there were no fuel or gas leaks after any engine change. As on almost any aircraft there will always be a leak, they used to say of the Lightnings and F-4 Phantom that if it did not leak, it was empty. To establish a datum for leaks on the TSR-2, Len would collect any leaking fluids in a container and the time it had taken to accumulate then, having measured the fluid, would have his datum point. Any further leakage would indicate there was perhaps a more serious leak than normal. During the preliminary shake-down tests gave Len the first opportunity to experience changing an engine out in the field. Due to the unserviceability of one of the fully instrumented engines, it was decided to replace it with one which had little in the way of test instrumentation. Although the task took a whole day, Len felt that with practice the time taken could be improved upon considerably.

The working hours were long prior to the first flight. Len and his team would start at 8 am and often work right through till late into the following night. After a day's ground running out at the area on Boscombe known as the 'Pear-Drop', the aircraft would be towed back to the hangar where the oncoming shift would rectify any problems found during the day. They would also carry out any modifications required by the BAC design offices. Once back at the hangar, Len would carry out a debrief of the day's work and schedule any changes required for the night-shift to action. It was found that work done during the night shift was completed speedily, simply because there were no interruptions by visiting VIP's and other people wanting to poke their noses in.

Len was impressed with the TSR-2 and with the men who worked on it. However, he was the first to point out certain shortcomings when it came to working on the aircraft especially when in the field. As is mentioned elsewhere, it was very much a specialised aircraft needing special tools and 'tall thin men with

# TSR-2 - PHOENIX OR FOLLY?

long arms' in order to reach areas deep within the airframe. The engine itself perhaps presented the best example of this type of problem. Before the engine could be removed, the gearbox and ancillary equipment had to be removed, and these were only accessible through very small covers on the airframe. This same type of problem had first been brought to light with the unusual design of the Lightning, which also required similar specialised tools and men of a similar stature to work within the airframe.

As the senior ground crew member Len was perhaps more aware of the problems that were affecting XR219 than anyone else at the time. The undercarriage appeared to be one of the most confounding problems, but signs that something was wrong had been apparent prior to the first flight. The aircraft was brought daily from the hanger to the 'Pear Drop', well away from the busy commotion of an active airfield. Although this distance was relatively short the aircraft was towed to this hard standing area by a tractor, Len had noticed that during these daily runs the tyres appeared to be subject to a large amount of wear and did in fact order a complete tyre change prior to the first flight, an indication that perhaps the toe-in on the front set of wheels on the bogie was excessive.

After the cancellation of TSR-2 Len continued working at Warton on the Lightning. In 1967 he became involved in the Saudi Arabia defence programme which was set up by BAC in Saudi itself. As part of the BAC team, Len spent some time in the country working on the Saudi versions of the Lightning before returning to Warton. Shortly after this he began preparing for the Jaguar and this also involved a period of time away from Warton at the French flight test centre working on French Jaguars. When Jaguar flight development began in Britain Len returned to Warton. The Tornado was to be the last development aircraft in which he would be involved before his retirement. Again this meant joining the team in Germany as part of the British detachment working with the German prototype before returning to Warton to begin work on the British prototype.

*The Tornado was another aircraft that both Bee and Jimmy were to become involved with. Bee as Director, Flight Operations for Panavia, and Jimmy, initially as Manager, Flight Operations and later as Director. Here seen through the lens of Flt. Lt. Mike Lumb of 14 Sqn, RAF Bruggen, five Tornado GR1's formate on a Victor Tanker during the 1991 Gulf War. (Rolls-Royce plc)*

# CHAPTER SIX
## The Proving Ground

During the initial stages of any new aircraft's flight development one man in particular becomes familar with the characteristics of the aircraft whereby he will attain a standard of knowledge and experience to pass on to those who will eventually assist him. The first critical flights on the TSR-2 were to be carried out by BAC Chief Project Pilot (TSR-2) and test programme director, Roland Beamont. When he had established the basic controllability and safety of the aircraft and gained sufficient experience on the type then further pilots would be brought into the programme. As Deputy Project Pilot (TSR-2) at the time, Jimmy Dell was involved in the initial stages of the TSR-2 project with Roland Beamont, and would be the second pilot to fly the prototype.

Both of these pilots were involved at the design phases where discussions took place on varying aspects of the aircraft's systems and the cockpit layout. There was a need for them to attend various meetings, especially with the engine manufacturer, and Bee and Jimmy often took it in turns to attend. It had been hoped that both pilots would fly in the Vulcan test bed XA894, which was being used for the airborne engine development trials of the Mark 320 Olympus engine, and in this way acquire valuable experience of the engine handling qualities. But this was not to be and only Bee was to gain any experience on the 320 engine in his one and only flight in the Vulcan. Three months later the aircraft was totally destroyed by failure of the first Olympus 320 engine during a ground run. Bee and Jimmy also spent time at the Institute of Aviation Medicine where they became involved with the new Aircraft Equipment Assembly. The pilots and navigators of TSR-2 were to wear sophisticated integrated flying suits and safety harnesses along with specially designed flying helmets. Both suit and helmet were individually tailored. It was therefore important that, Jimmy as deputy project pilot, should become completely up to date with all the briefings Bee received right up until the first flight. Indeed had Bee been taken ill then Jimmy, as the deputy project pilot, would have immediately taken over to ensure continuity in the programme.

**Testing Begins.**
Vickers, being the lead contractor, had wanted the first flight, (which would have been to A&AEE Boscombe Down in Wiltshire), to be made from the Vickers airfield at Wisley which was just outside Brooklands. However calculations showed that the runway there would be too short should an emergency arise. Roland Beamont had suggested movement of the aircraft by road to Warton where the runway was much longer than Wisley and slightly wider, although neither runways were as suitable as the one at Boscombe. Warton, being a BAC main plant, would also have been the ideal design support and maintenance base had any serious problems arisen during the testing and taxying trials. But the decision was finally taken to send XR219 by road to Boscombe Down and so, in April, the aircraft was duly dismantled at Weybridge, loaded onto a transporter in sections and shipped to Boscombe. To accomodate the forthcoming trials a whole new test and support facility had to be set up at Boscombe Down before aircraft testing could begin.

A modern aircraft is comprised of many components and whilst it is possible to check many of these out on a bench or test-rig, the only proof it will really work is

## TSR-2 PHOENIX OR FOLLY?

*A unique shot of XR219 (background) and XR220 (foreground) in the weighbridge hangar at Boscombe Down. A whole test facility had been set up by BAC to cope with the TSR-2 programme. XR220 showing no signs of damage after being tipped off the trailer. (BAe)*

when all the pieces come together in the airframe. That time had now come for TSR-2. The airframe was unloaded from the transporter and the task of rebuilding XR219 began under the watchful eyes of Bill Eaves, Superintendant in charge of TSR-2 testing and servicing at Boscombe, and Len Dean. Reconstruction of the aircraft took approximately 3 months and was carried out in the newly built 'weights hangar'. This was Boscombe's purpose-built weighbridge facility with all ground testing services laid on. This facility had now been taken over by BAC for the TSR-2 and its team. Soon after the aircraft had been reconstructed, engine and systems testing began and continued right through until September. In 'The Olympus Story', many tales are told of incidents that affected this long programme. The first taxying trials started at the beginning of September and were carried out by Roland Beamont. It was during taxi runs 6 and 7 that Bee was to be proved right in his opinion that TSR-2 should not be flown from the small airfield at Wisley. He recalls in his book *'Phoenix into Ashes'* a problem that arose when testing the braking chute. The chute was to be deployed during a high speed taxi run at some 140 kts and when the pilot pulled the chute handle the chute itself deployed and 'candled'. With the end of the runway looming up fast Bee had no option but to brake hard in an attempt to bring the heavy aircraft to a halt before it went crashing through the boundary fence and across the busy main Salisbury road, possibly ending up in a field. It is here that the braking system on TSR-2 came into its own. Don Bowen who was in the back seat was looking at the brakes through the large periscope and could see pieces of red hot material dropping from the wheels during the heavy braking. Despite this the only casualty was a tyre that blew its core plug under the extreme pressure. On checking the throttles it was found they had reopened under the braking inertia and Bee had closed them sharply to reduce thrust. This emergency action put the throttle through the H.P cock detent and inadvertently into the closed position, which left the aircraft with no electrical power and therefore no power to feed the electric fans used to assist the cooling of the brakes. With the possibility of the brakes welding themselves to the discs, Bee quickly released them before the aircraft came to a standstill. A modification was introduced to the throttle gate before the next taxi test to ensure that this problem did not happen again.

XR219's first flight was on the afternoon of September 27th 1964 and this historic event is covered very well in a number of publications written by the test pilot of that momentous occasion, Roland Beamont. However, a brief look at the events that led up to it may help those not familar with what occurred.

# THE PROVING GROUND

*A great deal of ground testing was to be done on XR219 before its first flight. Engines, fuel and hydraulic systems were checked and double checked before the aircraft was allowed to fly. (BAe)*

Though the forthcoming General Election was increasing the pressure on the team to get the TSR-2 airborne, safety was still paramount, and to jeopardise the whole project and people's lives at this stage would have been totally irresponsible. Bee had already consulted with BSEL, who had given him all the detailed information regarding the condition of the flight engines. At the time these engines had not been cleared for flight as BSEL could not yet guarantee their integrity. It was up to Bee then to decide whether the risks involved were acceptable and also to assure the authorities that any deficiences, which at this stage were numerous, were to an acceptable standard and therefore would not jeopardise the safety of the crew, people on the ground or the aircraft itself. Not until all these conditions had been met would the first-flight receive official authorisation.

## First Flight.

The original test schedule had already imposed specific limitations on XR219 for its first flight and these were quite extensive. However, the current problems that affected the engines and undercarriage, involved even more limitations which were to restrict the flight even further. BSEL had strongly recommended that power used during the flight be no more than 97% in order to keep out of the range where there was a possibility of the L.P shaft 'resonance' occuring. But the climb-out would be done with the undercarriage down and would therefore require all the power available. This would be even more critical if there was an engine failure after take-off, for Bee would need all the power available from the remaining good engine in order to get the aircraft to a safe height. The time scale anticipated during that first flight was expected to have been around 50 minutes, with the aircraft exploring a considerable number of flight parameters. The redefined schedule reduced this flight to two wide circuits of the Boscombe Down field which was expected to take around 15 minutes. The pilot would be exploring basic handling characteristics within the undercarriage 'Down' placard speed of 250 kts. However, if Bee encountered any problems during the first circuit, he would have to bring the aircraft in straight away instead of overshooting and continuing for a second circuit.

## TSR-2 PHOENIX OR FOLLY?

In the morning Bee made two high speed taxi runs to confirm the tail chute reliability, especially after the recent problems. Then, at approximately 3:15 in the afternoon, he pointed XR219 down runway 24 and committed the TSR-2 to the air. During the flight the undercarriage was left down and locked as tests on the gear had not yet met the necessary standard for a retraction. Problems were occuring during the retraction sequence and so far the planned ten successful undercarriage retraction cycles had not been met. A further five successful retractions were required to be done on the aircraft itself whilst on jacks inside the hangar. The auxilary intakes were also locked in the open position so even had the wheels been retracted, these doors would have stayed open.

During the flight John Carrodus flew chase in Canberra WD937 with a photographer and Jimmy Dell also flew chase with BAC photographer, John Whittacker, in T-4 Lightning XM968. This was the job Jimmy would do for all of the flights made by Bee at Boscombe Down as well as the transfer flight to Warton. He would also fly chase for Don Knight's single flight at Boscombe. Both Jimmy Dell and John Carrodus had already taken off in their respective aircraft. As Bee

*TSR-2 XR219 takes to the air for the first time on September 27th 1964. (BAe)*

*Jimmy Dell flying in a T4 Lightning nudges in close to XR219 during the first flight to allow company photographer, John Whittacker to take this excellent shot.(BAe)*

## THE PROVING GROUND

began the take-off run, Jimmy in the Lightning, positioned his aircraft to the rear and port of XR219. When TSR-2 became airborne Jimmy noticed large vortices coming off the wing tips. His first thought was that the wing tanks were venting fuel and he prepared to take avoiding action. However, he quickly realised this was not so, but merely condensation in the expected wing leading edge vortex at the point where the wing tip angled down - an understandable mistake as the wing tank vents were in that area. Similar vortices were also evident at the outer part of the blown flaps when in operation and looked quite spectacular as they both streamed away from the aircraft.

The duration of the flight was approximately fifteen minutes as per the schedule, and explored the basic handling and control of the aircraft during the two circuits. Even though it was a short flight it was, according to Bee, very obvious that this was 'a pilot's aeroplane all the way'. The control and stability of the airframe, even without stability augmentation, was quite exceptional, taking into account the philosophy of its design, with its high performance and low level capability. The take-off had brought no drama, however the landing caused some concern when the wheels touched down. In the words of the test pilot, 'All hell broke loose', and for a few seconds Bee came close to losing control of the aeroplane. In the landing stage, during a normal touchdown with this type of undercarriage configuration, the rear bogie wheels touch down slightly before the front due to the angle at which the wheels lie in relation to the fuselage. Then as the aircraft settles down, so the front wheels come into contact with the runway. It appears that as soon as the front wheels touched down a powerful vibration was set up and although this only lasted a few seconds it was still sufficient to cause disorientation at the cockpit. As we have seen earlier when studying the undercarriage retraction sequence, it was a complex movement. At first it was thought that the problem may have been due to the amount of toe-in on the front wheels, which had set up the initial vibration. This problem occured and was investigated on each of the flights at Boscombe, and a more in-depth study of the problem was planned for when the aircraft had been flown to Warton. No other major problems had been uncovered during the flight itself and for the next few weeks XR219 was to be laid up in order to complete the

Pilot and navigator of the TSR-2 first flight. Don and Bee enjoy a refreshing drink after the first flight.
*(R.P.B Collection)*

undercarriage testing and to allow the Olympus engines to be replaced with two units with interim modifications to eliminate the L.P. shaft resonance problem. This actually delayed the second flight, which had originally been scheduled for the end of October, until the end of December.

With temporary modified engines fitted, Bee flew XR219 for flight number 2 on 31st December 1964. During the flight a new problem was to manifest itself, again concerning the engines. At first Bee believed it to be a reccurrence of the resonance problem, for there was a vibration coming from one of the engines. Bee had already reduced power on both engines out of the suspect range which was between 97% and 100% Nh as a precautionary measure, but because the vibration was now starting to impair his vision he reduced the power on the No 1 engine even further and it was not until he had reduced it to below 87% that the vibration became tolerable. By experimenting with the throttles he found the worst levels to be between 87% and 96% on No 1 engine only. This was obviously a new problem and the flight was curtailed. A thorough inspection of the offending engine was made but no obvious cause was found, and so on January 2nd 1965 on flight 3 Bee, again with Don Bowen flying as observer, tried again. It soon became obvious that the excessive vibration was still there. Bee decided to abort the flight and did a tight circuit bringing XR219 in for its shortest flight, a total of only 8 minutes, because of the very real possibility of at least one of the engines disintegrating. It was found that an incorrectly manufactured pressurising spring in a low-pressure fuel pump was setting up the offending vibration and after it had been replaced, Bee was satisfied with the engines.

This problem highlighted a characteristic of the airframe which, although not causing any immediate problems, could still quite possibly affect the future testing and development of the airframe and its components. The vibration caused by the faulty low-pressure pump was felt by the pilot sitting in the front cockpit, but the navigator in his cockpit only feet behind him remained unaffected by any change in vibration levels of the airframe. It was later discovered that the long fuselage actually had various points where it was particularly susceptible to these high vibration levels. It was most fortunate that the pilot's cockpit area was one, so the vibration was quickly detected.

Another quirk had been found in the hydraulic system and this had first come to light during the ground tests on the hydraulic rig. The problem was still obvious

Up until flight 10 XR219 flew all of its initial flights with the undercarriage down. This restricted the aircraft to the undercarriage down placard speed of 250 kts. *(BAe)*

## THE PROVING GROUND

during runs made in the TSR-2 flight simulator at Weybridge. It was found that the hydraulic system appeared to be what can only be described as 'lumpy'. During taxying trials with XR219, the problem was still evident and giving some concern. However, Bee had declared that he was prepared to accept this for the first flight, but a modification should be sought straight away. It was found however, that as soon as the aircraft became airborne, this 'lumpiness' disappeared and the hydraulic system performed perfectly. It was only whilst the aircraft was on the ground that this symptom was evident.

As explained in the previous chapter Bristol Siddeley had been required, by the Ministry of Defence to fit a single channel warning system between the LP and HP shafts of each engine. These were linked to warning lights in the pilots cockpit. It was during Flight No 3 at the critical stage of the take-off that both of the red warning lamps came on. This subsequently proved to be a failure in the warning system itself.

### Problems with the Undercarriage.

Flight 5 on the 14th January was the third where there was to be an attempt at an undercarriage retraction and if one reads Bee's excellent accounts of this particular phase in the development of the TSR-2 in his book, *'Fighter Test Pilot'*, one can appreciate the problems and dangers these men had to face. Jimmy recalls this particular phase of the flight development very well and remembers a number of occasions when he had to inform Bee that 'such and such a wheel had either failed to retract' or that a particular bogie had failed to rotate on its beam.

These two dramatic shots show Bee landing XR219 after the bogies locked in the up position. The descent rate was measured at 6 inches per second. Such precision was achieved by the tremendous response of the TSR-2 flying controls. The photographs were taken from a high speed cine film, hence the poor quality.
(R.P.B Collection)

# TSR-2 PHOENIX OR FOLLY?

It was not until February 6th on flight 10 that TSR-2 flew in a clean configuration. Malfunctions and failures to retract occured on each of Bee's attempts on Flight 3 and 4, and then on Flight 5 the major problem occured. The 'UP' selection resulted in the main undercarriage bogies jamming in the vertical position from which they could not be moved. This was a fully justifiable 'Abandon' case but Bee decided to try a very smooth landing. As the loading of the aircraft came off the wing and onto the undercarriage at normal descent rate the impact of the wheels on the runway might have caused the bogies to rotate on their beams against the designed direction of rotation. This could have resulted in severe damage to the undercarriage which, without doubt, would have had dire consequences. Therefore Bee decided to use the exceptional precision of the TSR-2's taileron control to effect a landing so smooth the bogies would hopefully swing into their correct position. Bee gave Don Bowen the option to use his Martin-Baker ejection seat but he elected to stay on board! This was a courageous decision by him at a tense moment. The resultant landing was measured at the remarkable low rate of 6 inches per second vertical descent at touch down,This is even more remarkable when one considers that the normal vertical descent rate on TSR-2 was 10ft per second! The bogies rotated smoothly backwards into correct contact with the runway. This problem with the undercarriage was overcome by increasing the hydraulic rotational force on the main bogie. In other instances, where the pilot was getting false information in the indicator light system, the micro-switches were adjusted.

## Jimmy Dell's first flight.

The next flight was number 6 on 15th January and was to be Jimmy's first flight in the TSR-2. With Don Bowen in the back seat, this was a 24 minute familarisation flight, still with the undercarriage down. Jimmy announced he was extremely happy with the way things had gone, apart from the landing when he too experienced the severe undercarriage vibration. TSR-2 was one of the biggest aircraft Jimmy had flown and although he had taken over the controls in Britannias, VC10s and other large airliners, the TSR-2 was something totally different. He commented, *'the aircraft actually handled like a Lightning and one had to be very*

Jimmy in the T4 follows Bee down to the deck. Jimmy did all the chase flights from Boscombe and the transfer to Warton. Because of the confidence in TSR-2 chase flights were dropped shortly after XR219's arrival at Warton. *(BAe)*

# THE PROVING GROUND

*careful not to exceed the specified limits for the prototype. It was extremely easy to fly and for a large aeroplane, remarkably easy to land, it was also one of the most comfortable military aircraft I've had the pleasure of flying'.*

His second sortie was on 23rd of January, flight number 8, and still with the undercarriage not retractable, lasted a total of 27 minutes. This was an expansion of the previous familarisation flight and explored further the flight envelope he had already flown. During the flight the only problem was a reccurence of the reheat pump staying in stall. This happened after Jimmy had selected reheat in-flight. This same problem was to affect Bee's flight when he selected reheat during take-off for flight 9.

For flight 10 on 6th February, the intention was to investigate undercarriage functioning following the latest, and to date, the most comprehensive system overhaul which had included increasing the bogie rotation hydraulic pressure. As on previous flights the planned schedule provided for an expansion of the indicated airspeed envelope when the undercarriage had been retracted and lowered again satisfactorily, with the cycle being repeated three times. Crewed by Roland Beamont and Don Bowen, the undercarriage was cycled smoothly and satisfactorily soon after takeoff with correct sequencing of the warning lights. Then, after two satisfactory repeats, Bee confirmed over the radio his intention to investigate the high speed envelope

### XR219 fly's Clean.

Conditions were far from suitable with low cloud and mist, and overweather flying had not been cleared because of the undercarriage restrictions; so the tests were set up at 1500 to 2000ft keeping below cloud. At 50 knot increments Bee checked stability and responses in pitch, roll and yaw up to the scheduled 400 kts, and then as all was going well he decided to carry on up to the 500 kts current 'flutter' precautionary limit, (this would be lifted when the aeroplane had been cleared in ground resonance tests).

The day's tests went perfectly and at 500 kts the TSR-2 was found to handle as crisply and precisely as a fighter, to such an extent that Bee brought it back down the Boscombe runway at low level and 480 kts; a sight which gave the onlooking engineers and ground staff much pleasure after what had been a long and frustrating winter. Now the TSR-2 was in business and the more advanced testing could proceed.

On 6th February 1965 XR219 on flight 10 flew clean for the first time. This was an important phase in the testing of TSR-2. This shot shows the strong votices coming from the anhedral section of the wing. *(BAe)*

# TSR-2 PHOENIX OR FOLLY?

In view of the questionable reliability of the Olympus engines, it was necessary to evaluate single engine handling. One engine would not be actually shutdown, but throttled back to idle in order to simulate failure. As the engines in the TSR-2 were close together within the fuselage, only a minimal need to retrim the aircraft for single engine flying was expected. During Flight 11 on February 8th, Jimmy explored this particular aspect of the flight envelope by checking control and stability in XR219, first with one engine idling and then the other.

It was equally important that the aircraft stability trials be performed quickly to establish any basic flaws in the aerodynamic design of the airframe. One particular test involved trimming the aircraft for straight and level flight then, with hands off, the pilot would give the rudder pedal a kick. This would disturb the aircraft directionally and demonstrate the roll/yaw ratio (Dutch Rolls). Jimmy recalls the occasion when he was flying chase in the T-4 and Bee was flying XR219 for the roll/yaw tests that particular day. Bee had trimmed the aircraft up for straight and level flight and proceeded to kick the rudder pedal. The aircraft yawed slightly, but the amount of roll was phenomenal, so much in fact that Jimmy thought XR219 was going to roll completely, but Bee gathered it all together and continued on his way. TSR-2 was a long aircraft with only a small wing span and therefore had an expected high roll/yaw ratio. Bee reported that with familarisation it would pose no problem.

## Don Knight's first flight.

On February 10th, Don Knight, (with Peter Moneypenny), flew flight 12, lasting some 38 minutes, but it was to end rather dramatically. For his first flight in the TSR-2 it was a cold, windy, miserable grey day. This was to be a familarisation flight for Don within the flight envelope previously tested. The flight itself went well considering the weather conditions at the time. The landing phase began according to the flight plan and the descent rate was within the design parameters

Compare this shot of the landing configuration with the shots of Bee landing with the bogies locked up. Even at this stage there was still a vibration problem with the undercarriage during the initial landing phase. *(BAe)*

# THE PROVING GROUND

laid down for that aircraft. Touchdown though was somewhat firmer than he intended and resulted in quite a severe oscillation which was somewhat difficult to contain. The problem was aggravated by nosewheel steering over-sensitivity which had already been the subject of criticism. The aircraft was brought to a halt on the damp runway and the engines shutdown. Various service and emergency vehicles attended XR219 and after a brief inspection of the undercarriage, it was deemed safe to tow it back to the hangar. At the time Don felt the aircraft could have been taxied back under its own power, but examination of the undercarriage revealed a fractured forestay jack on the port undercarriage. This would be subject to a redesign to meet the specified rate of descent requirement.

**Transfer to Warton.**
The aircraft was laid up for necessary repairs to the undercarriage. Then, on February 16th on flight 13, Jimmy flew his last sortie out of Boscombe Down in XR219. It was necessary that the cleared flight envelope be extended to 30,000ft/0.9M to cover conditions for the forthcoming transfer flight from Boscombe to Warton. This was also to be Don Bowen's last flight in the TSR-2, as the majority of navigator/observer development work on XR219 would now be moved to Warton where it would be handled by Peter Moneypenny and Brian McCann.

During flight 13 the flight plan was to reach 30,000 ft and then cruise with the engines running at around 90% power. They flew for approximately 45 minutes across the local countryside assessing the aircraft and felt quite confident there would be few problems under similar conditions during the journey to Warton.

The flight to Warton, flight number 14, was to be flown by Roland Beamont with Peter Moneypenny assisting as navigator/observer. Jimmy would be flying chase as usual in the T-4 Lightning, again accompanied by a photographer. XR219 was lacking certain essential items of equipment for this flight and therefore specific weather conditions were to be imposed. The actual equipment shortfalls are covered in Chapter 2, but the main concern was that the engine anti-icing system had not been cleared for flight and was therefore missing from XR219.

February 21st, the day scheduled for the transfer, dawned cold, grey and gloomy. Low cloud hung over Boscombe airfield and the weathermen had said there was a probability of high icing index in the 8/8ths cloud cover along the route to Warton. It was apparent the transfer would not be taking place, mainly because of the lack of anti-icing equipment on the engine. The sceptics promptly seized on this opportunity to belittle the TSR-2, saying 'it was such a super aircraft, it could not fly unless the pilot could see the ground'. Such talk was, of course totally irresponsible and unfounded. Bee's decision had been based on the regulations laid down in Air Publication 67, which governs the flying of prototype aircraft. The regulation dictates the conditions under which any prototype aircraft can fly and one particular regulation covering this type of flying, stipulates the ability of the pilot to be able to see the ground whilst flying a prototype aircraft with limited navigational aids. As we have already seen, such aids in XR219 were indeed very limited. The following day the weather conditions had improved marginally for the journey to take place, and the crew began their pre-flight checks.

The journey to Warton lasted just 41 minutes and on the way Bee took the aircraft out over the Irish Sea (for supersonic investigation) in order that the sonic boom be kept away from populated land areas. XR219 and the chase T-4 headed out toward the start of Warton's test run 'Alpha' and when they arrived, Bee lit the afterburner on the port engine only, with Jimmy selecting reheat on both his Avons in the T-4. Much to his amazement XR219 just left the T-4 standing. This display of speed for such a large aircraft was quite outstanding and Jimmy calculated that 219 was doing about Mach 1.2 to his Mach 1.1. At Warton a large number of the workers had assembled on the airfield in order to see the TSR-2 for the first time, and Bee with one eye on the fuel gauge, gave a low fly-past before bringing the aircraft in for a copy-book landing.

XR219 in a clean configuration heading out towards the start point Alpha for the high speed transfer to Warton. Bee was flying Peter Moneypenny navigator/observer and Jimmy in the T4 is flying chase. *(BAe)*

XR219 rolls to a halt on Lancashire ground. Jimmy's T4 is in the background. The flight development could really begin now. *(BAe)*

# THE PROVING GROUND

*A pleased pilot and navigator, Bee and Peter after delivering XR219 to Warton. (R.P.B Collection)*

*Another view of XR219s arrival at Warton. The number of staff that turned out to meet it is evident. (BAe)*

## Warton and beyond.

XR219's arrival at Warton resulted as planned in a rapid increase in the pace of testing and developing the aircraft. As part of the strategy the responsibility of flight testing was handed over by Bee to Jimmy after Flight 16. Bee would be devoting more of his time to his job as Director, Flight Operations on the Warton Board. It was essential that the aircraft type attain its full flight envelope clearance and to achieve this it was important that a solution for the problem concerning the undercarriage vibration was found quickly. At this point no further detailed undercarriage testing was scheduled. Confidence in the aircraft was now extremely high and the familar pattern of TSR-2 being accompanied by a 'chase' aircraft was now dropped and TSR-2 flew with less limitations on longer, unescorted sorties.

During the following weeks Jimmy became involved in assessing techniques for landing to eliminate or at least reduce the vibration to some degree. He tried various types of approaches and differing methods of touchdown. This resulted in varying descent rates that either put the aircraft down very gently or really banged it onto the deck, but each time the excessive vibrations reccured. Finally a foam path

# TSR-2 PHOENIX OR FOLLY?

was sprayed on the runway by the emergency services to provide reduced coefficient of friction at the point of touch down and so reduce spin-up drag. This did reduce the level of vibration, but not enough and it was still evident that the problem required urgent design attention.

Just two days after its arrival at Warton, XR219 was flying again, this time in the hands of Jimmy Dell and Peter Moneypenny for flight 15. This was the longest duration flight XR219 was to carry out and lasted approximately 70 minutes. Flight 16 on the following day was to be quite significant for a number of reasons. Firstly it was to be Roland Beamont's last flight in the TSR-2 and secondly, this was the flight when he actually rolled the aircraft. Accompanied by Peter Moneypenny in the back seat, Bee was to expand the flight limits and had been doing Dutch Rolls and partial rolls through 60° and 90° and then South of Warton, he rolled XR219 completely through 360°. Jimmy, flying chase, was somewhat surprised at this manoeuvre as it had not been filed in the flight test schedule. However Bee had been satisfied with the response and damping of the lateral tests and had decided, as was his practice, to make a further careful exploration on his own initiative. This was normal practice at Warton at the discretion of the project test pilot.

The main aim now was to expand the cleared flight envelope of the aircraft further in an attempt to regain some of the time lost due to the initial problems with the engine and undercarriage. At the same time further attention was given to the vibration problem with the undercarriage and the majority of sorties included a number of 'touch-and-go' landings. On March 8th Jimmy Dell with Brian McCann flew flight 17, a 52 minute sortie in the morning. This was Brian's first sortie in the TSR-2 and during the flight they did a number of 'touch-and-go' landings at varying rates of descent to see if any particular sequence could overcome the landing vibration. Also, during this flight, Jimmy investigated simulated single engine flight control. During the afternoon sortie the flight was curtailed after 35 minutes when a serious leak in the 'fueldraulic' system was discovered. The same problem was to bring to a halt flight number 19 on March 11th after a further 33 minutes. The next day, March 12th, Jimmy Dell with Peter Moneypenny managed a 46 minute flight. This again expanded the flight envelope, but also took in a number of touch-and-go landings.

In an attempt to find a cure for the undercarriage vibration the emergency services sprayed foam on the runway at Warton. Seen here is Jimmy Dell landing XR219 on the foam. Although there was a slight reduction in the level of vibration, it was essential a cure for the problem was found. ( R.P.B Collection)

# THE PROVING GROUND

Flight 20 had brought a successful conclusion to the first phase of the flying programme. XR219 was now to be laid up for necessary modifications in order to continue with the next phase. It was decided to fit a temporary fixed tie strut between the oleo leg and the rear bogie. Although the engineers believed this would eliminate the oscillation in the undercarriage the temporary strut would not allow the undercarriage to retract. On March 26th Jimmy, with Brian, flew flights 21 in the morning and 22 in the afternoon. These consisted of touch-and-go landings to discover if the tie strut worked. The results suggested that the problem had been cured. As a result of this instructions were issued for an operational undercarriage tie strut to be manufactured that would permit normal retraction. This important modification would never be test flown.

To avoid delaying the next phase of the programme, the flying was to continue, although this would be restricted to the slow speed envelope only, due to the undercarriage having to be left down. Flight 23 was to be flown by Don Knight with Peter Moneypenny in the back seat. This was to be the last flight for them both in the TSR-2 and was of approximately 34 minutes duration. Although this was Don's last flight, he was however to prepare for a third flight, Flight 25 in early April. With Peter Moneypenny as navigator/observer, he had taxied out and was in the final stages of pre-flight checks at the end of the runway and was running the engines up just prior to releasing the brakes, when suddenly the whole cockpit came alive with alarms and indications on the pilot's central panel warning of a problem. There was a sudden loss of hydraulic pressure in one of the 4 systems. At the same time over the R.T came an urgent warning from the observation caravan situated at the end of the runway. They had noticed fluid pumping out at the rear of the aircraft. Don immediately aborted the take-off and shut the aircraft down. XR219 was towed back to the hanger for inspection where it was discovered that the hydraulic fin jack had fractured allowing the hydraulic fluid to escape.

On March 31 1965, Jimmy Dell, with Brian McCann as navigator/observer, flew XR219 on a 32 minute flight out of Warton before the aircraft was to be laid up for major modifications to various components, the undercarriage in particular. The flight investigated the roll/yaw gearing assessment. The aircraft's flight controls were at the rear end of the aircraft, and therefore when the pilot initiated a turn at slow speed, a certain amount of input would be needed from the fin. Eventually this would be taken care of by the autopilot, however, the exact amount of input required had to be assessed by the test pilot. This was to be the very last flight of a TSR-2 and is so recorded in Jimmy's flight log-book. After this flight Jimmy, with Peter Moneypenny, returned to Boscombe Down to prepare XR220 for its initial trials and first flight.

In all XR219 had flown a total of 24 sorties, which had gathered a great deal of important data. All the taxying trials and the first 10 critical test sorties had been flown by Roland Beamont and had clocked 4 hrs 3 minutes actual flying time. Jimmy Dell had flown 12, of the later, longer sorties which had given him approximately 7 hrs 50 minutes in the aircraft type. Don Knight had only flown 2 flights, one on familarisation from Boscombe and his second from Warton, for a total time of 1 hr 10 minutes. When XR219 was eventually grounded it had spent only 13 hrs 3 minutes in the air, but more than twice that 'running' time which included all taxying and engine trials.

The initial trials work on XR219 has already been covered, but a lot would have depended on solutions having been found to both the engine and undercarriage problems before moving on to the next stage of development. The aircraft itself was

*XR220 passing through Andover en route to Boscombe Down. Later that day a state of gloom hung over members of the project when XR220 was dumped on its side on the Boscombe tarmac. (Andover Post)*

scheduled to continue flight envelope clearance well into 1966, and in April XR219 was expected to be handed over to the Boscombe Down pilots for an initial 'preview' assessment on the aircraft. Then in the latter part of May and early June, it had been allocated to canopy jettison trials, with a complete canopy jettison marked for February 1967. After some 150 hours flying and having completed the scheduled programme the aircraft was then to have been used for possible flight development in the overall flight test programme.

**Setback at Boscombe.**
XR220's arrival at Boscombe Down on September 9th 1964, caused quite a stir. With another aircraft joining the test flying it was felt that with two aircraft in the air there would be a marked feeling of progress. That feeling was to be shattered by the end of the day. The journey to Boscombe had been relatively uneventful. The convoy with a strong presence of police, had made good progress. The vehicle towing the trailer with XR220's fuselage on board pulled up outside the hangar allocated to TSR-2. After a brief discussion it was decided the fuselage would be put in the hangar before being unloaded the following day. The vehicle would reverse the trailer into the hangar and then uncouple it. The vehicle turned sharp right, but the trailer did not have a sufficient turning circle to match the towing vehicle and when its front wheels locked the trailer tipped over depositing its load onto the tarmac. The fuselage and trailer lay on its starboard side. The police moved in quickly and whisked the driver away, whilst a survey of the damage was made.

Eventually the fuselage was righted and moved into the hangar to be cloaked in

# THE PROVING GROUND

*Disaster at Boscombe. From eye witness accounts, Phil Kingham has captured the moment when XR220 was thrown onto its side outside the hangar at Boscombe. Had this accident not happened the history books would have spoken of the two TSR-2s that flew. (Phil Kingham)*

covers. The fuselage was rejigged to determine whether it had twisted. It was found however, that the damage was not as bad as had at first been envisaged. The brunt of the fall was taken by the starboard taileron stub and, obviously, various panels were damaged. This most unfortunate accident was to badly affect the ground running trials on XR220 and put back any hopes of getting two aircraft into the sky for many months. Following this serious setback it was clear that had more thought and consideration been put into the mode and quality of transport then this whole incident could have been avoided and history would have recounted the story of the two TSR-2s that flew. It is also worth noting that had such an accident happened at Warton or Wisley the delay in getting the aircraft repaired may not have been so long.

XR220 had been scheduled to join the flight development programme at the beginning of March 1965 when it would have made its maiden flight. However, due to the accident, the aircraft remained unserviceable until the end of that month. By the middle of March XR220 had been repaired sufficiently to start engine trials prior to its first flight. This was a tribute not only to the aircraft's structure in that no major damage had occured to the airframe during the accident, but also to the men who had worked so hard to get the aircraft prepared.

It was planned that Jimmy Dell and Peter Moneypenny would carry out a number of taxi runs, if there were no problems then XR220 would fly that day, April 6th 1965. The actual taxi run was delayed because a problem was found with a fuel pump during pre-flight engine runs that morning. Towards midday the Chief Ground Engineer elected to have the pump replaced, purely from a safety point of view. During the lunch period, Jimmy and Peter visited a Social Club in Amesbury for lunch, away from the hassle of all the VIP's and dignitaries. Jimmy rang the field to ascertain if XR220 was cleared for flight. Told it was not, he and Peter began a game of ten-pin bowls. The club steward switched on the television to listen to the budget speech. Jimmy was conscious of the voice saying *'...Twopence on cigarettes, sixpence on beer...* and then that the government had decided, in the interests of the country, to cancel the whole TSR-2 project. They rushed back to Boscombe Down in the hope of getting XR220 airborne, but on arrival they were told by the Crew Chief and senior officials that the aircraft, although ready to fly, had been grounded because of the recent communication. The aircraft was then removed to the hangar to await further instructions while the political wranglings continued.

# TSR-2 PHOENIX OR FOLLY?

At this stage the aircraft had demonstrated to good effect the true concept of its design. Even without the sophisticated systems destined for the project, TSR-2 had once again proved that with the right resources the British aviation industry was capable of accomplishing the ultimate in aviation design and manufacturing. The excellent stability and precision control of the aircraft had surprised many, and even without stability augmentation the ride was extremely comfortable in both the high speed/high level, and high speed/low level/rough air flight envelopes. Although the wing had been designed as a compromise, it already appeared very efficient and would cause the pilot no problems in the event of an automatic systems failure. On many occasions both Bee and Jimmy had said that the TSR-2 is as easy to fly as the best Mach 2 fighters. Such confidence enabled the flight test crew to push the aircraft, confidently and safely beyond the limits laid down in the test schedule, and the rapid flight test progress had helped to regain some of the time lost.

The avionics and navigational equipment being developed in conjunction with the airframe was also proving to be at least as good as, and in certain instances, even better than had originally been anticipated. Indeed, the numerous companies manufacturing this equipment went on to demonstrate the viability of their products by gaining significant overseas orders after the project was cancelled.

The following is from a summary report written by Roland Beamont prior to the cancellation and circulated to MoD, Air Ministry, Farnborough, Boscombe Down and the Boards of the BAC and BSEL companies.

*This then is the TSR-2; a brilliantly successful pilots' aeroplane. Potentially one of the most effective military aircraft of all time, TSR-2 was specified by, designed, built and flight tested for the Royal Air Force. That it would achieve all that is expected of it is beyond serious doubt, and this is to the lasting credit of the team that designed and built it.*

Literally within a few days XXR220 was ready to fly, but those days were to prove costly. With only hours before take-off the government brought a sudden halt to the programme. The aircraft was retained at Boscombe Down to confirm test results on the Olympus engine for Concorde. *(BAe)*

# CHAPTER SEVEN
## Behind Closed Doors

*The idiot who praises with enthusiastic tone,*
*all centuries but this,*
*and every country but his own.*

Gilbert and Sullivan.

XR219 arrives at Warton. Even though the programme had made up a great deal of lost time, the threat of cancellation was ever present. *(BAe)*

Politically, the TSR-2 project was frought with many problems, some perhaps even more complex than the very mechanisms being built into the aircraft itself! The project, though born under a Conservative government, was nevertheless still a target for those who were against the morality of such weapons, and also those who felt that this type of weapon was an unnecessary drain on the nation's resources. It was also a fact that being born under the Conservatives, did not necessarily make it immune from the threat of cancellation. In common with other areas of national expenditure at the time, the cost of defence was increasing and naturally any measures that could be introduced to limit such costs would have been extremely welcome. So when Peter Thorneycroft, now Lord Thorneycroft, announced orders for three new types of aircraft, there were those that condemned them as a typical Conservative waste of money, naturally with the loudest voices of condemnation coming from the opposition.

# TSR-2 PHOENIX OR FOLLY?

As the TSR-2 project progressed, so the campaigns against it grew. Predictably a project of this calibre will always attract opponents with conflicting interests: some with political influence, some merely wanting to use such projects as launching platforms for their own careers, and some just wanting to better their station in life. At the time the TSR-2 was the project that was to mould the foundations of the British Aircraft Corporation and therefore appeared to be the only current military project that stood any significant chance of survival in the present climate, but at what cost?

**Election Buildup.**
The Conservatives had been in power for 13 years and it was evident the wind of change was coming. In 1963 the Party had been rocked by the 'Profumo' affair, when it was discovered the then Secretary of State for War, John Profumo was having a close liason with a lady of questionable morals, whilst the lady was at the same time, having a similar affair with an officer from the Russian Embassy, and thus it was felt that the security of the nation had been seriously compromised along with Mr Profumo.

During 1964 both political parties launched their election campaigns with the country still not totally convinced that either party would be able to make any inroads into the problems of the time. Harold Wilson, as leader of the Labour Party, visited various constituencies in an effort to win support, but his awareness of the real state of the country had to be based purely on assessment, rather than first hand knowledge. He knew however that things were not particularly good and he began making promises that could probably only succeed as a result of strong measures, based on a series of drastic cutbacks.

Whilst visiting Preston, Harold Wilson was continually bombarded with questions from people, especially those employed in the aviation industry, seeking some form of assurance about the future of the TSR-2. After Wilson's visit he was quoted in the Press as saying, 'that if the aircraft worked and the country could afford it we could have it'. In the run up to the election it was said that pamphlets had been seen in Preston denying that the TSR-2 would be cancelled by a Labour government. Later when the local Labour candidate for Preston South, Peter Mahon had won the seat he disclaimed any knowledge of such a leaflet! Indeed the wind of change was certainly blowing over Preston, with very few taking comfort in the words of Harold Wilson.

**A Labour Victory ?**
As some had predicted, Labour did indeed win a closely fought General Election. With only a 77% turn out, the election was dubbed 'the one the Conservatives had lost', rather than 'the one Labour had won'. Labour's majority was only four and even this was reduced to three the following month, when Patrick Gordon Walker failed to win a local By-Election. Even so, it was sufficient to convince those involved on the TSR-2 project that the government review, so strongly hinted at in pre-election speeches, was imminent and would soon establish the fate of the programme.

The Labour Party have never been noted for successful policies on the economy or defence, and many felt that yet again perhaps quite rightly, this would be no exception. It would appear that all those years in opposition had done little to improve matters and that too many of the newly elected members set to grace the front benches, including Harold Wilson himself, had little knowledge on the

# BEHIND CLOSED DOORS

various aspects of defence other than to conclude that overrall it was a grossly expensive commodity. It was obvious then that the Cabinet would have to rely on the expert knowledge of those who knew and understood such matters. It was perhaps to be expected that these 'advisers' would have their own preferences regarding their own fields of expertise with the defence policy. But nevertheless it was anticipated that a cross-section of these opinions would at least provide the government with a broadly based and unbiased viewpoint. It was ironic though that many of these so called 'advisers' appeared to have been either influenced by the American institutions or hold an allegiance to a particular service.

### Those Against. Lord Louis Mountbatten.

One particularly eminent figure with influence was Lord Louis Mountbatten. In 1959 he became Chief of the Defence Staffs, only the second person to hold this post. This was a position from which he was to bring a great deal of influence on certain military projects being considered by defence chiefs. As Chief of the Defence Staffs, Mountbatten, like his predecessor William Dickson, should have displayed total impartiality in matters concerning all three armed services. There were naturally many conflicts of interest notably between the Air Force and the Navy, and many believed Mountbatten abused his position in order to let the Navy have its way.

There are a number of instances where the Navy appeared to get preferential treatment as a result of the influence by Mountbatten. In 1962, the British government were negotiating with America to purchase a new missile system which would give the Air Force a strategic nuclear capability. America had cancelled Skybolt which in effect had left Britain without a missile system to negotiate with. Although Mountbatten himself was not present at Nassau, he had little cause for concern as his right hand man, Solly Zuckerman, had accompanied the Prime Minister, and was able to advise and influence his decision to buy Polaris.

Had Macmillan secured an agreement on Skybolt, then Mountbatten had been fully prepared to confront the Air Force as to their reason for procuring such a system. As it was intervention was not necessary as the government went ahead and ordered Polaris. The affair was seen as another Mountbatten - Zuckerman winning combination.

Lord Louis Mountbatten of Burma. A great naval man, but as Chief of the Defence Staffs his judgement appears to have been clouded by his allegiance to his navy, and became detremental to the other services. (TWMOC)

The assistant Chief of Staffs, Air Marshal Sir Christopher Hartley, was also at Nassau and perhaps the only person there capable of making any assessment on weapons that could be used in conjunction with the current bombers and TSR-2. It appears he was excluded from all negotiations which effectively eliminated any contributions by the Air Force.

When it came to the question of TSR-2, Mountbatten did not remain impartial, openly declaring a preference for the Navy's Buccaneer. Mountbatten was convinced that the Buccaneer, with some slight modifications, would ideally suite the RAF. He also felt that if the Air Force went ahead with the TSR-2, it would not only virtually bankrupt the defence budget, but the Air Force would still be left with an aircraft which fell short of its design requirements.

If the RAF were to accept the Buccaneer, not only would it help the Navy considerably with the development costs of the aircraft, but it would also ensure there were sufficient funds in the defence budget for the proposed new 'through carriers' for the Navy. Mountbatten turned to his close friend and ally, Solly Zuckerman for help. Zuckerman at the time was the Chief Scientific adviser to the Minister of Defence. As the CDS, Mountbatten had to at least show impartiality, but felt he could, through Zuckerman, perhaps still influence the Minister of Defence. He wrote a brief for Zuckerman suggesting that the contents be put forward to Harold Watkinson, the Minister of Defence.

The following is a note that was attached to the brief, and is taken from Philip Ziegler's Biography of Lord Mountbatten.

*'This is the first occasion on which your action is absolutely vital to the Country's Defence Policy, and to save the Minister from making a ghastly mistake. You know why I can't help you in Public. It is NOT moral cowardice but fear that my usefulness as Chairman would be seriously impaired. BURN THIS'.*

Zuckerman was unable to help and a few days later Mountbatten decided to write to Watkinson direct. In his letter he acknowledged that his own outspoken views against the TSR-2 could undermine his credibility as an impartial Chairman of the Chiefs of Staff, and therefore he was unwilling to repeat these views in public and risk the wrath of the Chief of Air Staff. He then spelt out his arguments for the Buccaneer and those against the TSR-2, pointing out that if he were Chief of Air Staff he would have gone for a modified Buccaneer or the Barnes-Wallis variable geometry aircraft. Mountbatten did not want to fight the Air Ministry and cause ill feeling with his Chiefs of Staff, but neither did he want to see millions of pounds being spent on an aircraft he felt would never get into service. However, the fighting did not stop here.

Later on in 1962, Mountbatten was to adversly effect the outcome of a campaign to sell TSR-2 to Australia, and this is studied later on. The consequences of his campaign were to cause him problems in the coming years. When the Labour Party came to power Mountbatten was able to renew his policy, this time with a Defence Minister who was more amenable to Mountbatten's proposals.

**Sir Solly Zuckerman.**

Sir Solly Zuckerman is a South African who had been the Chief Scientific Advisor to the Minister of Defence since 1960. He was a zoologist and a leading authority on glandular behaviour in monkeys. Although his publications are very few, especially those on defence and aviation, there are a number of papers on the relationship between man and apes. He appeared to have shunned seeking knowledge of Britain's aviation industry, preferring instead, he said, "to keep an open mind and free from any particular influence". However, this certainly did not appear to be the case when he visited the United States. As a friend of Robert McNamara, America's Defence Secretary, Zuckerman had the opportunity of visiting a number of different aircraft manufacturers, and did so as their guest. From these visits his knowledge of the aviation industry must have grown, allowing him the opportunity unfairly to compare the two industries. As his knowledge was based on the American way of working the British side of course never gained favour. It appears that he was in favour of leaving Britain's major aircraft work to other countries, such as the United States. They would do all the research and development then allow Britain to purchase the finished product off the shelf, hopefully at a much reduced price, thereby saving millions of pounds. This would leave British designers and engineers to do more important and useful work. A

*Lightning production at English Electric was typical of that used by the British aircraft industry at the time. Zukerman was convinced that the American method would be far more efficient. (BAe)*

report, written by Zuckerman in 1963 on Defence Research and Development, had very little good to say on the subject, and although his conclusions are hazy, he does make the point that he considered the TSR-2 an unnecessary expense. Harold Watkinson was completely aware of Zuckerman's opposition to the TSR-2.

Zuckerman had been an ally of Lord Mountbatten when they both served in Combined Operations, and they were to work together again in 1959 when Mountbatten succeeded William Dickson as the Chief of the Defence Staff. Once in office, Mountbatten was to change drastically the workings within Whitehall. He was keen to see the standardisation of the needs of the three services, instead of allowing them to pursue their individual needs without reference to each other. Mountbatten became the sole voice for the Chiefs of Staff, instead of each Chief reporting to the Minister or Prime Minister direct. He held regular meetings with his Chiefs of Staff inviting Zuckerman along and soon the Mountbatten - Zuckerman combination became a powerful force in the corridors of Whitehall. There seems little doubt, that the influence of these two powerful men had a great bearing on the outcome of the TSR-2 project.

**Richard Worcester.**
There were other not so eminent people who used their voices and position to pour scorn on the TSR-2 project. One such person was Richard Worcester, described as an 'Aviation Expert'. He has written a number of papers on the subject of aviation and in 1966 wrote a book entitled *'Roots of British Air Policy'*. His main influence appeared to come from the two years he spent in the United States working on the magazine *'American Aviation'*. His time here was taken up in producing a weekly news sheet called *'Aviation Studies'*, funded from America, and prior to this he had worked in Britain on the staff of the magazine *'Aeroplane'*. Some of his claims contained within the pages of his book are quite extensive, and in fact one may wonder why his name has not been more prominent if, as he claims, he was able to

influence the Labour government on its defence policy to the extent that he did during the period 1964/65. A number of his articles on aviation also appeared in *The Observer*, a newspaper which attempted to influence public thinking by casting doubts on Britain's ability to sustain a nuclear deterrent. Such claims would bring into doubt the neccessity for an aeroplane with the capabilities of TSR-2 which, according to the same paper, was being built by an industry that was growing fat on government funds.

His book is perhaps the most revealing on his vision of Britain's air policy. It is packed with names, facts and figures and decries the industry and various government departments, with the RAE coming in for special mention. His criticism of the industry is based on its alleged inefficiency and lack of foresight, particularly in aircraft design. Some of his points are valid, but when someone writes a book that questions the integrity of a particular industry to the extent that Worcester has, then they are bound to uncover a number of inefficiencies.

On the subject of TSR-2, Worcester quotes Harold Wilson, saying that, *'the TSR-2 would not have suited the changing needs'*. This statement is not clarified and fails to answer what is actually meant by the 'changing needs', nor does he say what the new needs might be. Still in awe of Harold Wilson, Worcester goes on to portray the Labour Prime Minister as the only government leader since Lloyd George's time who had any real grasp of aviation, accusing other leaders of delegating the responsibility for air policy to their air and war departments. I have always believed that the reason why the elected party forms a Cabinet is so that the individual ministers can be responsible for the numerous tasks it must perform. Worcester's book appears to be nothing more than an attempt to justify a confused Labour policy on aviation and defence, and attempts to point out the advantages of going to America for all our aviation needs.

During the time of the TSR-2 Worcester was known to hold left wing views. He also appeared to associate himself closely with a number of policy making decisions made by the Labour Party, although this association was denied by the party at the time. In 1965 he did actually present a paper to the Royal Aeronautical Society at his own request. This paper turned out to be nothing more than a personal denunciation of the British aircraft industry. His conclusion was tantamount to

Apart from the Canberra, only the Meteor and the experimental Short Sperrin (below) had the engines on the wing...

# BEHIND CLOSED DOORS

*... totally contrary to Richard Worcesters statement in his book on British aircraft design.
(Shorts - Rolls-Royce)*

suggesting that Britain should become virtually just another state of America who should handle the defence of the West, on the grounds that they appeared to be the only country able to handle a potential conflict on such a scale. His paper also failed even to acknowledge the pioneering achievements of the British industry, and apparently he felt the energies put into it would be better channelled into other industries, such as micro-electronics, at which we do well.

Worcester's praise of the Labour government's actions appear to be misguided insomuch as his quote regarding Labours 'real grasp of aviation' intimates a good knowledge of the subject. If this was so then why did the industry not see a clearer policy on aviation when the Socialists came to power. In truth this appears to be nothing more than an attempt by Mr Worcester to justify his association with Labour party policies. The grasp he speaks of appears to be nothing more than Labour's decision to seek assistance from America in the form of aviation products, something Richard Worcester had so strongly intimated in his book.

**Mary Goldring.**
Another critic was Mary Goldring, at the time, air correspondent for *The Economist*. Miss Goldring was also a keen supporter of the American aviation industry and felt very strongly that the British aircraft industry should, in some way, be abandoned and that Britain should look to America for its aviation requirements. This policy alone would release workers and technicians in the industry to do 'other more important jobs', although what these jobs were was never

# TSR-2 PHOENIX OR FOLLY?

quite explained. She also considered the aviation industry to be unreliable and accused it of letting the Services down too many times over the years through its inferior products and inability to deliver goods on time.

Just after the war the industry was seeking constantly to build new and better aircraft, and often spent large amounts of its own money in putting manpower to work on new technology only for the government to cancel at every opportunity. Therefore it was not so much the industry that was letting the services down, but more the government in refusing to back it. In January 1965 when the Labour government announced the cancellation of the P1154 and HS681, Mary Goldring was apparently very disappointed that the TSR-2 had not also been included in the cancellation package. Miss Goldring felt that had the government cancelled the TSR-2 there and then, this alone would have saved some £30 million which could have been better spent on new, smaller aircraft projects she felt would be more suitable for the RAF. What the 'new smaller aircraft' would do is uncertain, but what is clear is that it was highly unlikely that the Labour government would have spent ANY money saved from the TSR-2 project on new aviation products for the RAF.

The fact that Miss Goldring felt the RAF should have smaller 'Baby Bombers', illustrated that she appeared to have no understanding of the role the TSR-2 was to play. The Operational Requirement called for the aircraft to have a range of 1,000 miles, to ensure that TSR-2 could penetrate enemy territory and attack and disable specific targets such as airfields, radar and radio stations, fuel dumps and so on. Smaller bombers would have neither the range nor the capability to achieve this because of the lack of space for fuel, but would also lack the sophisticated radars and on-board computer systems necessary to pinpoint their targets with anything like the required accuracy. In January 1965, Miss Goldring admitted that she did not really know what the TSR-2 was supposed to do, which from someone who had just spent the past months appearing on television virtually every week expounding her theories on why and when the TSR-2 should be cancelled, was to say the least, quite an admission. One popular misconception Miss Goldring shared with large numbers of the general public was the belief that the huge cost of the TSR-2 was due to its nuclear capability. This confirms how badly misguided she was, in believing that because an aircraft had a nuclear role it must cost more. The widespread destruction caused by a nuclear weapon meant that the placing of such a device did

It appears Mary Goldring's assesment of the TSR-2 was flawed by her ignorance of the aircraft's true capability. *(BAe)*

not require the same high level of accuracy as did an ordinary H.E bomb. Therefore the sophisticated devices to deliver such weapons were attributable to the ordinary bomb.

A passage in Stephen Hastings book, *Murder of TSR-2,* sums up Mary Goldring's attitude to the British aviation industry, *'... to crab, to criticise, to cancel and to praise our competitors'*. This type of destructive tactic appears to be common practice with the media in this country, but rarely does one see evidence of this kind of behaviour in the States. In America the Press are more involved which gives them the ability to make better assessments and promote its country's achievements.

Miss Goldring also had misgivings regarding the Concorde and poured scorn on this project. During a BBC radio interview on the Concorde she was heard to say, with a chilling disregard for humanity - the best thing that could happen would be for the Concorde to crash on its first flight. A recent admission about her earlier misgivings on the Concorde perhaps demonstrates that those first years spent decrying the British aircraft industry may well have been misguided. It is possible that her earlier efforts into researching the TSR-2 may have been thwarted by the veil of secrecy surrounding the project, and that all her assumptions were based on secondhand information. It can only be hoped that had she come fully to understand the logistics of such a project, her assessment would have been much more favourable than it was.

**The Australian Interest.**
Selling the TSR-2 to a foreign power had always been a distinct possibility and had an export contract have been won, it would certainly have helped to relieve the pressures being put on the project at the time. In retrospect the affair appears to be one of the most mismanaged sales drives the British government has yet had a hand in. As early as 1959, Jeffery Quill, Military Sales Manager BAC, and Roland Beamont, gave a presentation on the TSR-2 to the Australian MoD and Air Force. A further series of follow-up presentations on the technical aspects of the aircraft were also given during the following year. Then, in March 1962, Sir George Edwards, then Chairman of BAC, visited Australia as a follow-up to the 1959 campaign, and found the response to these presentations most favourable. The Australian government had been coming under increased pressure, mainly from the press, to invest more on defence, especially with the growing unrest in the Far East. The TSR-2 appeared to be the ideal aircraft, although other types were also under consideration. These were the American MacDonnell F4 Phantom, the North American Vigilante, and the French Dassault Mirage. The F-111, although it was discussed at the time, was not yet a serious contender.

BAC continued the pressure on the Australians and kept them informed of the details and progress of the programme. It was in the March of 1963 that Australia's defence minister, Sir Frederick Scherger, visited Paris for a meeting. During this trip he also visited Britain but was unable to visit BAC or other departments involved with the TSR-2. He was however able to attend a meeting in London where he met with Lord Mountbatten and Sir Solly Zuckerman What actually took place at this meeting is not documented, but Scherger left the meeting with serious doubts about the TSR-2. During lunch with the then Air Minister, Julian Amery, Scherger had voiced his concern about Mountbatten's opposition to the TSR-2, although Zuckerman was quick to point out that nothing was said at this meeting to cause Scherger to have any doubts.

# TSR-2 PHOENIX OR FOLLY?

It was rumoured that Mountbatten had six special cards made, five depicting Buccaneers and one the TSR-2, and at meetings he would casually slap them on the table saying, *"Five of these for the price of one of those"*. Eventually Watkinson instructed Mountbatten to cease his attacks on the TSR-2 as he felt the Buccaneer could not meet the Air Forces requirements.

It was following this visit that the Australian interest suddenly began to wane. Thus it was vital that BAC should convince the Australian parliament that the aircraft would still be built, and continued their campaign even in the light of the recent meeting. The Australians requested the loan of a squadron of Vulcan bombers as an interim measure in maintaining their strategic capability. The request was relayed to the British Prime Minister by BAC but the government failed to respond, Harold Macmillan maintaining that the request should come direct to them from the Australian government. When the request was finally acknowledged by the British government, there were certain reservations. The main concern was Australia's intention to man the British aircraft with Australian pilots. This of course meant that the aircraft would come under the jurisdiction of the Australian government. Britain would have prefered to base a squadron of Vulcans in Australia with British pilots, but this was unacceptable to Australia and they duly turned the offer down. They were concerned that in the event of hostilities occurring, there would be serious confusion over which government would give the order to send the aircraft into any conflict.

This offer from the British government to loan the Vulcan bombers to Australia would no doubt have gone some way to ensure, not just the future of the TSR-2, but would also have secured an early export customer for BAC.

In 1963 Australia ordered the F-111, saying that the cost of the American plane was considerably less than that of the the TSR-2. This was perhaps one of the most

F-111As on the production line at Fort Worth. Australia chose the F-111 claiming it would be cheaper than TSR-2. They received their aircraft 10 years late and paid nearly treble the original price. *(General Dynamics)*

# BEHIND CLOSED DOORS

publicly damaging statements made on the subject. From that point on arguments raged both in the House and in the Press as to the viability of such a plane. Such arguments also highlighted the large amounts of money being spent on the aircraft. It is highly probable that this was a decision Australia would regret. The F-111s eventually cost them $344 million against an original estimate of $140 million, with the aircraft being delivered 10 years late. At the time Australia could have had the TSR-2 at a unit cost of £2.1 million each.

**Problems for the Government.**
The long period of Conservative rule had left the Labour government with a considerable number of problems, in particular the state of the country's finances. The government's biggest worry was the £400 million trade deficit and the pressure it was putting on the pound. The various ministries were also seeking finances in order to carry out their election promises. Their concern on defence seemed centred on the three main RAF programmes, these were the HS681, the Hawker P1154 and the TSR-2.

The HS681 was a transport plane intended as a replacement for the Argosy, Beverley and Hastings. Its origins go back to the Armstrong Whitworth Company in Coventry before its amalgamation as part of the Hawker Siddeley Group. It was designed as a STOL medium transport aircraft and had been put forward as part of the support vehicle for the TSR-2. Its main function however was to carry 60 paratroopers with full kit at a speed of 325 knots. One of the main requirements had been for a large loading door in the side of the aircraft and this was one of the reasons why the C130 Hercules had not been considered. This problem appears not to have been insurmountable, with the manufacturers satisfying the customer's requirements in principle. The power plants were to be the Rolls-Royce RB142 Medway engines which had an unusual system of exhausting, each engine having twin rear mounted deflectors with a rotating cascade, which had been designed to give it its short take-off capability.

*An artists impression of the HS681, cancelled in favour of the American C-130 Hercules.
(Ian Frimston)*

## TSR-2 PHOENIX OR FOLLY?

The P1154 on the other hand was a proposed supersonic V/STOL aircraft from the Hawker camp based around the P1127 Kestrel/Harrier. It had originally been conceived by Peter Thorneycroft following the idea of Robert McNamara that a common airframe could be used for both the Air Force and Navy. At the time there appeared to be serious differences over what exactly the two services wanted from such an aircraft. The Navy in particular wanted two engines, twin seats, folding wings, nose and tail, a totally different radar system to the one required by the RAF and a whole host of complex equipment. Unfortunately the Air Forces's needs were virtually the opposite. What the Navy wanted the Air Force did not, and what the Navy considered a primary role, the Air Force considered only as a secondary requirement. These sort of complex differences certainly did not help the new project. The P1154 had actually been chosen by NATO to fulfill the Nato Basic Military Requirement, (NBMR3) role. However due to the French objections it never reached that stage and because of NATO's lack of funding for the aircraft, the whole project was dropped.

**Healey takes over Defence.**
Denis Healey, as Minister of Defence, was to consult his Chief of the Defence Staffs, Lord Mounbatten and the Chief Scientific Advisor, Sir Solly Zuckerman. Various matters needed to be discussed, and no doubt the subject of the TSR-2 would have been mentioned. Although no actual conversations are recorded, it is almost certain that when the question of the plane arose, there can be little doubt as to what was said on the subject. With both Mountbatten and Zuckerman having grave doubts as to the credibility of TSR-2, their views on the matter, together with the new confidential material available to Healey, would have a strong influence any decisions to be made.

**Plans of Cancellation.**
George Wigg, the Paymaster General, and Harold Wilson appeared to spend many hours together at Chequers discussing the options available to them regarding the forthcoming policy on expenditure. Defence of course was one of the major topics during these discussions. It was also around this period that Richard Worcester

The Lockheed C-130 Hercules. The RAF modified some of its aircraft by extending the fuselage which enhanced the capacity of the aircraft. Shown here is an original version (lead aircraft) and one of the modified types. *(Lockheed Aircraft)*

# BEHIND CLOSED DOORS

claimed to have had the ear of George Wigg, apparently being able to furnish Wigg with exclusive details on the technical performance of the TSR-2. As a matter of course it was deemed necessary to review, in particular, all current aviation projects, including the Anglo-French Concorde. Towards the end of October the French had been informed that the British government were seeking to discuss the future of the supersonic project. It may be remembered that Julian Amery, under a Conservative government, had already committed Britain solidly to negotiations with the French and a binding treaty had been signed, so understandably the French were adamant in refusing to allow the aircraft to be cancelled. Had Britain insisted on this course of action, the consequences could have quite easily proved to be most embarrassing, with Britain possibly being taken to the High Court. The results of such an action would no doubt have left us without an aeroplane, but would still have left us to contribute a tremendous amount of effort and money into a project with a now somewhat dented reputation.

### Concorde in Jeopardy.

As Concorde's development grew, so the anti-Concorde hate campaigns in the US gathered momentum. Ostensibly the main objections to Concorde were that the engine noise would create a serious health hazard and endanger buildings within the flight path. Such fears were groundless as TSR-2 itself had demonstrated during its development flights. Any hopes America had of building a similar airliner had been dashed by a lack of Congressional support, and it was felt that the anti Concorde campaigns in America would be instrumental in Concorde's failure to become an instant world beater.

### Further Pressures to Cancel.

The Treasury of course were also to play their part in the affair of the TSR-2 and could be included among those seeking to cancel such prestigious programmes. Indeed, they had already expressed their concern that such projects were a huge drain on public money. The workings of the Treasury, like any other government department, are so complex that to the man in the street they seem to be shrouded in an aura of mystery. The Treasury does however bear a tremendous responsibility in being the purse of the nation, so to speak.

Concorde. The worlds first (and so far, only) successful, supersonic passenger airliner. Its worldwide success appeared to hinge on the acceptance of America.
(Rolls-Royce)

# TSR-2 PHOENIX OR FOLLY?

In order to carry out that responsibility effectively, the Treasury would have had to agree a contract between themselves, the Air Ministry and the manufacturer. This would be for work to be carried out by BAC for and on behalf of the Air Ministry. The Treasury would then be responsible for releasing funds for work done annually. At the end of each year the Treasury would review the project and could either withold funds or even terminate the contract. Toward the end a review of the project was made every 6 months and it was only Julian Amery's tenacious attitude that prevented the Treasury from cancelling there and then. The biggest problem emanated from tenders that were initially submitted for approval. Obviously for a manufacturer to be successful in his bid, a low priced tender would be submitted. Once the work began the price would sometimes begin to spiral out of all proportion, leaving the government to be faced with the dilemma of whether to continue with or cancel the contract.

**First Defence Review.**
In November 1964 a defence review was held with Harold Wilson, Roy Jenkins and Denis Healey. Wilson had already suffered the embarrassment of having to give consent for a deal on arms to South Africa, although it was to bring in much needed money. The deal, which involved the sale of 16 Blackburn Buccaneer aircraft to South Africa for £20 million, had been first negotiated under a Conservative government, but now in its final stages of completion, still needed the final seal of approval from the new Labour government, despite its direct contravention of the embargo on arms sales to South Africa.

On the home front however, Healey having carried out a preliminary defence review, was pressing for all the current key military aircraft projects to be cancelled and for orders to be placed for American aircraft. Following this meeting, although no definite decision was reached regarding the future of the HS681 and P1154, it was obvious that cancellation was imminent. However, assurances were given to BAC that the TSR-2's future was secure until at least June 1965, when the programme would be subject to a more stringent review.

**Fickle Remarks.**
The reasons and excuses for cancellation were changing with the days. Condemnation of the project ranged from frivolous to outrageous, in many cases the accuser having very limited knowledge of the subject. The member of the House who could perhaps give a true account of the project was himself openly condeming the programme. A chance remark made by Healey during a luncheon demonstrated his contempt. Casually turning to Harold Wilson, he was heard to say, *"Oh by the way did you hear that a TSR-2 wing broke at Farnborough today?"* This comment taken on its own sounds most alarming to say the least, but when put in its proper context with all the facts, the correct meaning becomes much clearer.

As with any new aircraft, many of the components are normally tested to destruction and in this particular case a TSR-2 main wing had been on a fatigue test for some time in the laboratory at RAE Farnborough. As Denis Healey quite rightly said, 'it broke', but had done so only after standing up to stresses far beyond its original design specification and in fact had passed the structural fatigue test comfortably. The statement itself virtually typifies the unwarranted recklessness that politicians had at the time for this important project. Roy Jenkins, as the Air Minister responsible for the RAE, was annoyed that he personally had not been

# BEHIND CLOSED DOORS

XR219 on finals over Thruxton. It seems the Labour Government refused to recognise the potential of the British aircraft. *(BAe)*

informed of this event. However when he returned to his office he was able to discover the truth of what had happened. Although extremely unhappy with the situation, there was little he could now do. Far too many ministers had heard the story and any attempt to explain the truth would no doubt have fallen on deaf ears.

### BAC Fights Back.
On December 18th 1964, BAC invited certain air correspondents to a special assembly at Pall Mall. The intention was to put an end to some of the public speculation regarding the TSR-2 and at the same time put right some of the misconceptions regarding the capabilities and roles of the aircraft. This was perhaps the first occasion that BAC had openly made an attempt to clear up the controversy that surrounded the project. Those gathered heard a brief description of the aircraft and some of its equipment, together with an explanation of the roles intended for the TSR-2. During the meeting there was no attempt to discuss the rising costs, and the subject was never mentioned.

### Wilson goes to America.
During that same month, Harold Wilson with Patrick Gordon Walker, Denis Healey, the Chief of Defence Staff (Mountbatten) and members of the defence and Foreign Office, visited America for a meeting with Lyndon Johnson, the newly re-elected American President, along with other members of the American government.

The many problems the Labour leader faced during this visit were compounded by a deep mistrust the American's had of the new Socialist government. They feared that under Labour socialism would become rife in Britain and that the government would fail to respond positively to any threat of hostilities shown by eastern block countries. It was also reputed that Wilson's personal reputation in America had become severly dented through unsupported accusations of links with the Russian KGB.

The British and American leaders were to discuss a number of subjects, not least foreign policy and defence. It was believed however that the main reason for the visit was to discuss Britain's current economic crisis. Continual runs on the pound meant that the government were coming under increasing pressure to devalue. The pound had stood at the same value since 1949, $2:40, but devaluation now would mean huge cuts in public spending and a loss of confidence in Britain's standing as one of the major money markets.

The possibility of Harold Wilson requesting a loan from the IMF had been made clear by him in a Cabinet meeting on the 11th December, when he stated categorically, that while he personally had made no such request, America being very understanding of Britain's position, was ready to assist in any way they could. It was, however, very clear that something on the subject had been discussed, because in the following February the IMF extended the British government's credit to the value of £3,000 million. This American understanding of Britain's financial position had no doubt been formed as a result of Sir Eric Roll's visit to Washington just after the election victory in October 1964. As a Treasury official, Sir Eric had recently been appointed the Permanent Under-Secretary to the Department of Economical Affairs and had been sent to Washington in order to pave the way for the delegation the following December. He took with him a draft of the government's forthcoming statement on the economy. Part of this contained references to Britain's huge expenditure on low priority projects, mentioning in particular the Anglo-French Concorde and their desire to be rid of them.

**Healey talks Defence.**
Whilst Wilson had been talking to Johnson, Denis Healey had been in discussion with Robert McNamara, and the two defence men appeared to get on very well together. The main topic they discussed was Britain's alternative weapons system, and it may well have been during the 5 hour trip the two men took to Omaha, that they discussed the merits of the F-111 against those of the TSR-2. It was evident from Denis Healey's debrief to the Cabinet on his return, that he and McNamara had discussed a wide range of issues on defence and weapons. There were indications that Healey had agreed to Britain and America pooling their resources, specifically in the area of research and development of weapons. This would not only ultimately reduce the cost of R&D but, theoretically, would also reduce the eventual cost of weapons to both countries. Such collaborative work outside Europe would also lead to an effective reduction in military costs, specifically in Hong Kong and the Persian Gulf. This would enable Britain to perform certain duties for the alliance that it would otherwise be unable to do.

This sharing of R&D would certainly have been welcome in America, as the amount the US was spending on military aircraft development was phenomenal, with the XB70 alone costing some £286,000 per flight. Not only development on new aircraft, but also on defence itself. The wing of the B-52 bomber had to be strengthened because of a known weakness. The cost of this single programme was to be more than the price Britain paid for its V-Force. Such programmes illustrate just how great America's development programme costs were compared to those of Britain's.

**America's Concerns.**
The desirabilty of British aircraft had been shown to very good effect with the English Electric Canberra. In August 1950, the company had demonstrated their aircraft to a select committee from the U.S.A. As a consequence the Canberra was chosen to attend a fly-off in Washington the following January. The results of the

*TSR-2 during its first flight. America were deeply concerned that British technology would again prove to be far more formidable than its own. (BAe)*

fly-off culminated in the Canberra being selected by the U.S. Air Force as its new medium range intruder. However, because of American policy insisting that the USA would not be dependant on any foreign country for its arms material, the Canberra had to be built in the States. This job was taken on successfully by the Glenn Martin Aircraft Company, which went on to build over 400 various derivatives of the Canberra, called the B-57.

The TSR-2 had that same *'Made in Britain'* stamp all over it and the advanced technology carried within it was second to none. No doubt the Americans again saw this as an increasing threat, especially to their export markets. The British aviation industry continued to be a thorn in the side of the American industry. This had been demonstrated by Australia and the eagerness of their government to acquire the super bomber.

Allegedy America felt that by sanctioning Britain's IMF application would be tantamount to subsidising the very industry it had most to fear. Therefore the shallow promises made to Britain of financial support and subsidised F-111's, appeared a small price to pay for the cancellation of the TSR-2. There was even a promise of 'offset' trade to try and eradicate some of the costs involved in the aircraft purchase and at the same time, help reduce Britain's balance of payments. The cancellation of the TSR-2 would revive foreign interest in the F-111, even though it had only just made its maiden flight, and a tremendous amount of development work was still needed to get the aircraft flying properly.

**Concerns at Whitehall.**

The F-111 was in trouble prejudicing McNamara's reputation, as after all it was his brain-child. Congress badly needed some reassurance that the plane was as good as the original expectations. Therefore securing an early order with an overseas power, the British in particular, would help alleviate the current situation. The American Defence Secretary was eager to see his baby become a success, and with Britain in such a vulnerable position, this was an ideal opportunity he could not miss.

# TSR-2 PHOENIX OR FOLLY?

The Air Staff were becoming increasingly concerned at the events that were leading up to the eventual cancellation of the TSR-2. Within Whitehall there was a strong feeling that a change was about to take place, with many of the Air Staff expressing confusion as to where to place their allegiance, and others even on the point of offering their resignation. They knew that the TSR-2 was the only aircraft capable of fulfilling the RAF's requirements, but at the same time, the powers that be appeared hell-bent on cancelling the project. Denis Healey wanted them to have the F-111 and he felt that if he could somehow placate them with promises of the American plane, this alone would certainly smooth one aspect of the cancellation process. This was not to be and in an effort to avoid any further embarrassment, Harold Wilson asked George Wigg if he would intervene. The investigative journalist Chapman Pincher, in his book, *Inside Story*, tells of when George Wigg and the Chief of the Air Staff Sir Charles Elworthy, flew from RAF Northolt to RAF Abingdon where they spent most of the day in talks. The exact nature of these talks is not known, but it was very obvious that Sir Charles had been told of the irrevocable decision to cancel the TSR-2 and what's more that if the Air Chiefs continued to cause trouble over the project, then the Air Force would not get the F-111 as promised either. From that moment on the mood within Whitehall changed. BAC also began to notice these changes, for it was now becoming increasingly difficult to liaise with the very people who had previously been the backbone of the project. Significantly this included many high ranking RAF personnel who had been involved with the project.

Another source of unrest and concern to the Cabinet came from the large engineering Unions. Their awareness of the threat of cancellation was becoming more obvious and they were beginning to organise various forms of action in what would be a vain attempt to fight the cancellation. In January 1965 a mass demonstration was held in London with over 10,000 workers from the aviation industry alone attending.

**Problems with the American Plane.**

The F-111 development programme was not without its setbacks, but these were kept behind closed doors. The aircraft was originally designed to meet an American requirement known as Specific Operational Requirement 183, (SOR183), issued in 1960. When John.F.Kennedy came to power he brought with him Robert McNamara, an ex Vice President of the Ford Motor Company, as his Secretary of Defence. McNamara was a very shrewd man, having taught management at the Harvard Business School before moving on to become the President of the giant Ford Motor Company. He was known as a hard man with controversial ideas and was intent on getting his own way. McNamara's keen views on cost effectiveness had led him to believe the American Defence department could be run with the same effectivness with which he had managed the Ford Motor Company. This McNamara engineered by bringing in what we now term 'whizz kids', young men similar to himself, to help him sculpture a more efficient department. It was a strategy that was to make him a very unpopular man, particularly with the old, established members.

When the SOR had been issued by the Air Force, the Navy were also looking for an aircraft to fulfill their Fleet Air Defence Requirement, (FADR). It was McNamara's keen mind that realised both needs could be fulfilled by one basic airframe. A number of aircraft manufacturers submitted their tenders for the new SOR183, but when the decision was made as to which manufacturer should get the

# BEHIND CLOSED DOORS

lucrative contract many felt that in having chosen General Dynamics, McNamara had made a grave mistake. Even the Air Force advised against the F-111, as did the Navy, but the Secretary of Defence brushed aside all the expert knowledge and advice and went ahead, ordering the F-111. It appears McNamara literally instructed the manufacturers on how to design the plane in his bid to achieve commonality between the two military aircraft requirements. This incident has been a major thorn in the side of the Defence department for many years, and continues to cause problems.

**Design problems Foreseen.**

B.O.Heath one of the senior designers on the TSR-2 at BAC Warton had seen a detailed, 3-view drawing of the F-111 and decided to write a paper on the aircraft. Some of his observations included concern at the problem of boundary-layer air building up along the short forward fuselage and causing problems for the engine when it reached the intakes. With the engines being positioned very close to the inlet mouth the disturbed airflow often caused the engine compressor to stall. Another point he made was that the main wing chord was in line with the tailplane chord. This, he felt, was undesirable because of the likelyhood of the two acting as one when the main wing was swept back to its maximum angle of 72.5°. A further potential problem was excessive base drag in the tail area. In many of these instances he was proved right, for during the development of the aircraft most of these problems were to plague General Dynamics.

Three men who made a significant contribution to the TSR-2. H.H.Gardner, Julian Amery M.P and F.W.Page. Taken at Boscombe Down shortly before first flight. *(Part of the R.P.Beamont collection.)*

# TSR-2 PHOENIX OR FOLLY?

**Plowden.**
In conjunction with his visit to the USA, Harold Wilson had also specifically requested the formation of a committee chaired by Lord Plowden, to report on the state of the British Aircraft Industry. Many believed the industry was using large amounts of government money and producing very little to show for it. One area exempt from this investigation was the electronics industry, which at the time accounted for one third of the cost of modern warplanes. The Plowden Committee gathered on the 9th of December and considered the future of the aircraft industry, whilst taking into account the current economic situation in Britain. The question was, could and should the country afford such an industry in that present economic climate, especially as allies such as America had large resources and aircraft stationed all over the world? The industry was already struggling hard to establish itself as a major manufacturer of civil and military aircraft, any further intervention by the findings of the committee would continue to harm the industry.

Those on the committee were :-

| | |
|---|---|
| Lord Plowden. | Chairman of Tube Investments. |
| Mr David Barron. | Managing Director of Shell International. |
| Sir Casper John. | Admiral of the Fleet. |
| Aubrey Jones. | Chairman of the Prices and Incomes Board. |
| Mr Fred Heyday. | Officer of the U.G.M.W. |
| Mr Christopher McMahon. | Advisor, Bank of England. |
| Sir William Renny. | Chairman of the Atomic Energy Authority. |
| Mr Austen Albu M.P. | Member for a short time and replaced by Dr John Cronin M.P. |
| Mr St John-Elstub. | Managing Director, Imperial Metals. |
| Mr Walter Tye. | Air Registration, Technical Advisor. |

Lord Plowden was very sceptical of the task that had been set out before him. He was at the outset concerned about the amount of change that had already been instigated within the industry by a government which had previously announced the cancellation of a number of projects. Plowden was beginning to feel there was no real industry to investigate and had therefore approached Harold Wilson with a view to being relieved of his duty. Wilson however, was equally convinced that such an investigation was needed and thus insisted that Lord Plowden continue to form his committee. This he did, but remained unconvinced that his findings would uncover anything new, or for that matter invoke any government action.

The prototype F-111 on a test flight from Fort Worth. The Labour Government were told this aircraft would be cheaper and far more effective than its own TSR-2. *(General Dynamics)*

# BEHIND CLOSED DOORS

*Plowden had recommended that Britain should persue collaborative projects with our European partners. The SEPCAT Jaguar was the first military programme to be produced under a joint venture between Britain and France (BAe)*

Aviation's largest institutions, the Royal Aeronautical Society and the Cranfield Society recommended Plowden form an individual body of technicians, men drawn from the industry, who were capable of collecting, analysing and evaluating the information required by the committee. For reasons unknown Plowden failed to heed these recommendations and instead he consulted the MoA, who promised to supply all the necessary material. As a consequence the Royal Aeronautical Society contacted the MoA, drawing attention to the possibility of bias should all the information be seen to come from a single source. The MoA failed to respond.

As expected the main findings of the committee merely uncovered mistakes already made by the industry in the past. Perhaps the most significant recommendation was to encourage Anglo-French joint ventures, something that had already been pioneered with Concorde. This point alone was to give the industry some encouragement to pursue projects that would be formulated under a multinational governing body. Its effect was to determine the way the industry survived and is looked at later on.

**Meetings with Manufacturers.**

During January 1965, Harold Wilson had dinner at Chequers with the heads of the aircraft industry in order that they should be fully informed as to the government's thinking regarding current aircraft projects, including those in the military sector. While the civil aircraft programmes, mainly Concorde, appeared to have the blessing of the government, all military projects were to undergo very strict reviews, due mainly to the ever increasing costs. Also raised was the question of price. It was evident that Wilson was trying to obtain a fixed price policy that would not only give the government price guarantees, but also include penalty clauses, should there be further delays in delivery. Although not specific, Wilson intimated that the Government were already considering an alternative to the TSR-2. Whether he was using this as a form of intimidation in order to get the industry to realistically revue its costs or, whether it was simply a signal that cancellation was iminent, can only be a matter of speculation.

Unable to obtain any fixed price Wilson consulted with Denis Healey in order to formulate a cancellation programme of the three major defence projects. He decided to announce the cancellation of the HS681 and the P1154 during a forthcoming debate on defence, but was apparently loath to announce the demise of the TSR-2 at the same time. In the meantime, money was still to be made available for the TSR-2 until the end came. This was to annoy a number of Cabinet members, Denis Healey in particular, who did not want to see desperately needed defence money being thrown into a project that was to all intents and purposes already dead and was to be scrapped in the very near future. Tony Benn was of the same opinion. If the project was to be cancelled, do it now as any delay of the inevitable would not only incur extra costs, but would also draw attention again to the plight of the large numbers of aircraft workers who would be made unemployed by such action in addition to those threatened with lay-offs by the forthcoming cancellation of the HS681 and the P1154.

**Cancellations are Announced.**
In February 1965, during a House of Commons Debate, Harold Wilson duly announced the cancellation of the HS681 and the P1154. These cancellations alone were expected to save the government in excess of £300 million over the next 10 years, but the cancellation of the HS681 was to cause the eventual closure of the Armstrong-Whitworth factory in Coventry. Wilson also made it clear that he was not satisfied with the present costs of the TSR-2 and the way they appeared to be spiralling. He called for an even stricter evaluation of the project, concentrating on those areas of the development which were still causing most concern. The engines in particular were singled out for investigation, but he also mentioned his concern over the possibility of extra money being required to continue with the electronics development. Although at the time this equipment had not flown in an actual TSR-2 airframe, XR221 had been fitted with full avionics equipment and was currently undergoing ground trials at Weybridge. Therefore, with 85 percent of this equipment having already satisfied the prerequisite needs, such concern appears to have been unfounded. The only funds that would have been required were those to ensure the natural development of this equipment, funds that were accounted for already.

To replace the P1154, the Navy was to have the American F4 Phantom, but powered with a reheated Rolls-Royce Spey engine. These aircraft would be backed up with a number of Harrier/P1127s. To replace the HS681, it was decided to order the American Lockheed C130 Hercules aircraft at a cost of around £725,000 each. This price was indicative of the inducement now being put on Britain to purchase American aircraft, for at the same time the Australian government, who were also buying the Hercules, were having to pay nearly double at around £1,300,000 each!

Between February and March there was a tremendous amount of speculation in the press and on television regarding the future of the TSR-2, for many of the papers had already assumed there was little doubt the aircraft would be cancelled and that it was merely a matter of days before such an announcement would be made. Cabinet ministers, not directly involved with the project, were at the time genuinely unaware that any positive decision had been made on the TSR-2, and all the speculation was certainly causing confusion. There were those who had made it clear they wanted the project cancelled, and of course did little to deter such stories. At the same time there were others equally concerned that the project was worth saving and they tried to dispel the Fleet Street gossip.

# BEHIND CLOSED DOORS

**The Cabinet Meets.**

The full Cabinet was summoned to a meeting on the last day of March 1965, at which it was intended to reach a decision on the whole future of the TSR-2. There is no doubt that this was the meeting that sealed the fate of the aircraft once and for all. Certain Cabinet members however, were convinced that the matter had already been decided upon and that this was merely a charade in an attempt to convince Cabinet members and the public alike, that the government was playing fair. However, the opinions and views of those members outside the 'select few', would have little or no bearing whatsoever on the forthcoming discussions. These feelings were based on the fact that Harold Wilson had spent virtually the first three months of his term in office, concentrating purely on defence. Ironically, amongst those in favour of the TSR-2 was the new housing minister Richard Crossman, a minister who in fact may have benefitted more from the cancellation of the TSR-2 than many others, as he required money for his new housing policy. Fred Peart, Minister of Agriculture, and William Ross, Secretary of State for Scotland, were also included in those wanting to keep the TSR-2.

James Callaghan had already made his views known some months previously and as Chancellor of the Exchequer, it was no surprise to learn that his prime reasons were based solely in the interests of saving money. Denis Healey of course, wanted the TSR-2 out of the way in order that the option he had negotiated on the F-111K with McNamara could be taken up without more delay. George Wigg, on the other hand was shown to be treading more carefully on the matter. He felt the Cabinet should not only defer the decision on the TSR-2 and or any proposed replacement, but also allow more time for a further review of the present situation to determine finally, whether Britain really needed an aircraft of this type, especially under the prevailing circumstances. At the time the only replacement for the TSR-2 being put forward for serious consideration was, of course, the F-111K, and although Roy Jenkins and George Brown appeared to be in general agreement that the TSR-2 should be cancelled, they felt that consideration ought to be given to a British aircraft as the eventual replacement.

Until now Harold Wilson had not expressed any clear opinion. He had, though, shown disapproval of Jim Callaghan's opinion when Callaghan had broached the subject back in February. Although the discussions continued until lunchtime, there seemed to have been little progress made and no clear decision was reached. It was not until the the Cabinet met again later on that night that the prospect of a new development became evident. As the members reassembled, it became apparent that some negotiations had taken place during the afternoon. It seemed Denis Healey had somehow set up an alliance with various members, giving some vague assurances that taking up the option on the F-111s carried no obligation to purchase. Whatever the truth was regarding the alliance is unsure, but it appears Healey's negotiations had ensured he would get the F-111. Harold Wilson had still not indicated how he personally felt on the subject, but was quite definite about the way his Cabinet was to vote, by first making each individual member clearly indicate his preference. Finally, the subsequent vote was split three ways, firstly for outright cancellation, secondly, cancellation but with an order for F-111s and Phantoms, and thirdly to continue with the TSR-2. Certain members of the Cabinet felt that work still outstanding on the aircraft certainly did not warrant the investment needed, and so as the vote was generally against continuing, it was agreed that the project would be cancelled and the order be placed for 50 F-111 aircraft. For reasons unknown Harold Wilson wanted the cancellation announcement delayed, so it was decided to announce it to the country as part of the forthcoming Budget proposal in April.

**Wilson visits France.**
Prior to the April Budget, the Labour leader visited France for a meeting with General DeGaulle. It was during this visit that the French President was informed of the decision by the Labour Cabinet to cancel the TSR-2 and to go ahead with the purchase of American aircraft. In Harold Wilson's book which covered his period in office from 1964-1970, reference was made to this meeting and of a conversation between himself and DeGaulle. DeGaulle was at great pains to illustrate to Wilson, 'that you, the British, were and had been for many years satellites of the Americans', pointing out our dependency on America for items like arms and computing. Wilson responded by highlighting the recent takeover of the French computing concern, Machine Bull, which had just been acquired by an American computer firm. In the event this was just a feeble attempt to point out that France too had a certain amount of dependancy on America. True the French did possess a number of the American F-100's, but in no way were they committed to America for aid regarding arms, and, what is more, the aviation industry in France continued to enjoy not only the confidence of its government but also a healthy export programme of its ever popular 'Mirage' fighters and multitudes of successful missiles.

On April 6th 1965 the TSR-2 programme was cancelled. At the time production was well under way on the next pre-production batch of aircraft. *(BAe)*

**April 6th 1965.**
The manner in which the news was broken to BAC seems quite controversial. A number of top executives, including Sir George Edwards, Lord Portal and Sir Reginald Verdon-Smith, were invited to the House of Commons on April 6th 1965 and were told of the government's decision to cancel the TSR-2 project. BAC's relationship with its workforce was good having been nurtured over the years so, realising the devastating effect this news would have on the workers, the BAC executives requested permission to inform their management before they might chance to hear it on the radio. This was refused, for the announcement was to be part of the Budget Speech which of course was still secret. Thus any attempt to communicate this totally disastrous news to the outside world, prior to the Chancellor of the Exchequer's speech, was completely forbidden, and the executives were unable even to leave the building until the official announcement had been broadcast.

Sir George Edwards. Along with Lord Portal, took BAC through its formative years. *(BAe)*

Accordingly, amid the talk of general increases in taxation on certain goods and personal incomes, James Callaghan went on to announce the decision of the government to cancel the supersonic bomber, TSR-2. He went on to say that this alone would, over the next ten years, save the British public some £450 million. After the Budget Speech Denis Healey rose in the House and explained to the members gathered, the reasons why the TSR-2 had been cancelled. The House listened, powerless to question him because of the tradition in the House that forbade them to do so on such an occasion.

Because of the implications of the cancellation some members of the House felt this was not the ideal time for the Minister of Defence to justify such a policy. The House was in uproar with some calling it an act of treachery and others a 'scurvy trick'. The Conservative defence spokesman, Christopher Soames tabled that a Censure motion be added to the Order Paper. Whatever the government's idea was in announcing the cancellation at such a time was fast becoming their undoing. Even during Denis Healey's speech there had been nearly 50 alleged 'points of order' and with the Censure Motion tabled by Mr Soames, this was proving to be one of the roughest rides Denis Healey had come across as the Minister of Defence.

**Budget Debate.**
On April 13th 1965 a government debate took place on the TSR-2, perhaps the first open debate on the subject. The defence secretary was to put forward the main reasons why the government had chosen the course of action they had, and at the same time, explain why they had gone to America for a replacement aircraft. The opposition were still in no mood to let Healey off the hook, for after only 25 minutes he was having difficulty in making himself heard above the noise coming from the opposition benches. One of Healey's comments that caused uproar was his comparison of Britain's aviation and that of other countries. Giving praise for the industry on a man-to-man basis, Healey went on to say that until now it had been governments which had made the wrong choice in consistently choosing the wrong sort of military projects for the size of the potential market. One therefore has to assume that the Labour government were to pave the way in ordering the right

# TSR-2 PHOENIX OR FOLLY?

Another view of the prototype F-111. *(General Dynamics)*

equipment to defend the country. This was a government speaking which so far had never been known for its clear policies on defence.

**Fickle Promises.**

Although at the time Denis Healey was praising the F-111 as the right aircraft, the government had neither a firm delivery date, or a firm price. When questioned on this matter Denis Healey was a little hesitant and mentioned that around 1968 the RAF would be receiving F-111s for training. On the subject of price he was quick to point out that the government had received a firmer price from the Americans for the F-111 than from BAC for the TSR-2. When specifically questioned on the subject of cost and numbers involved Denis Healey declined to commit himself. The opposition were concerned that the Socialists had sold the country out to the USA, merely to gain a little time and in fact would be leaving the whole mess, including the pound/dollar crisis, for a future government to sort out. The wranglings went on with the opposition attacking and the government defending, although somewhat rather weakly. Sir Lionel Heald, Conservative Member for Chertsey got it right when he said, he did not believe the RAF would get the TSR-2 or the F-111 and that the Prime Minister had led the Air Chiefs up the garden path. Not only it seems the Air Chiefs, but also the British public, including those whose very income was based on a stable aircraft industry as promised in 1960 with the formation of BAC.

**A Question of Cost.**

One of the key issues in the whole affair is the question of cost. It is not my intention here to examine in detail the complex subject, this I will do in the next chapter. However, in an attempt to show the significance of the cost I feel some attempt should be made to demonstrate some of the problems that faced the industry during the course of the programme. In earlier chapters an in depth study was made of the aircraft itself. This included a look at some of the more complex equipment that was to be fitted in the plane, but of course there were many highly technical components that have not been described in order that the reader should not become bogged down with too many in depth descriptions. It would have required a number of volumes to cover such detail, but that which I have included gives a basic appreciation of the aircraft and some of the roles it was designed to perform.

There have been many different figures quoted on the this subject. An average figure was calculated from as many sources as possible with the figures being based on a contract for 100 aircraft, although originally it was to be for 150 planes. It will be remembered that the government accepted an initial tender of £90 million from English Electric and Vickers for the cost of nine prototype aircraft with development work to take place over a period of 5 to 6 years. Following this, a development contract was issued to Vickers, as the lead contractor, in June 1959. The engine contract was treated separately and was therefore awarded direct to the engine supplier, Bristol Siddeley, in December 1959 based on a revised tender. The figure submitted by the engine company seems questionable as it appears relatively low, even for the amount of work that had to be carried out in order to modify the Olympus engine to fly at supersonic speeds. This was before any serious problems started to manifest themselves during the development.

Sir Reginald Verdon-Smith. Formally with Bristol Aircraft before joining the Board of BAC. Later in 1968, on Lord Portal's retirement he became Chairman of BAC. *(BAe)*

For some unknown reason, the issuing of the main contract for the aircraft was delayed and was not issued to BAC until October 1960. By this time the original estimate of £90 million for the development work had risen to £137 million. This gave BAC its first real indication of how the project was proceeding, for it was only at this stage, when the design teams had practically completed their detailed work, that a better assessment could be made and the building of the aircraft could begin.

By June 1962, negotiations had resumed, this time in order to issue a further contract for 11 pre-production aircraft and a year later a firm order was placed with BAC for these aircraft. However, the development costs were still rising with the blame being placed on a number of factors, in particular the apparent lack of control on outside contract work.

Bristol Siddeley were also in serious trouble with the engine and their estimates had increased considerably. This was largely due to the costs of the Vulcan test bed which had been written off during testing, being added to the development figure. So by the middle of the year, the cost estimate was tagged at around £190 million, with the in-service date slipping. By October 1963, the Ministry were entering into negotiations with BAC for the supply of 30 production aircraft. It was realised that a contract would have to be placed very soon in order to allow for the long-lead time on materials and parts. The following February this contract was signed and BAC went ahead ordering the necessary materials. By January 1964, the government had changed, and signs of discord were becoming quite obvious.

# TSR-2 PHOENIX OR FOLLY?

It was around this period that BAC were in a better position to give the Ministry a more accurate estimate of the eventual costs of the project, which was £670 million. This figure included the £130 million for engine costs, which had been given to BAC by the MoA. There was £65 million mainly for category 1 equipment, which had been ordered by the Ministry and £475 million for the airframes. Sir Richard Way, Permanent Secretary to the MoA, hinted to both BAC and Bristol that this figure was far in excess of what was acceptable to the government, indicating that a figure of £604 million was more in keeping with what the government had in mind and that BAC should review their estimates. BAC had to consider this request very seriously and proceeded to trim their estimates, including those of the expected profits. In this way they were able to reduce the figure by some £45 million. This was still not good enough, even though in the worst possible scenario BAC could stand to lose £9 million.

The crux of the matter appeared to rest on the government being responsible for all losses in excess of 5 percent. Sir George Edwards had worked out a formula in order that a compromise could be reached. This involved the government and BAC sharing any losses of up to 5 percent, BAC keeping profits up to 10 percent, and then between 10 and 15 percent of profits being shared equally between BAC and the government. Anything above 15 percent would be retained by the government. At this stage there seemed to be very few 'chances' left to take, as the project was well into the development phase and the work was now going very well. The engine manufacturer was already working on modified engines, so theoretically the

XR220 was ready to fly on the day the project was cancelled. *(BAe)*

# BEHIND CLOSED DOORS

### "FOR SALE"

### B.A.C. T.S.R.2. (XR.219)

This vehicle is new in every way and must surely be the finest example offered for sale anywhere, designed for the man who requires something different.

Styled and built to a progressive new formula, must be seen and driven for its immaculate condition to be appreciated.

This fabulous vehicle is not for the inexperienced driver, and is in excellent running order and exceptionally well cared for, serviced very efficiently since new, with history complete in every way.

In immaculate condition (no body rust), finished in high gloss off-white cellulose finish, very low milage, and with 3 careful drivers, suspension suspect, tenacious road holding, tyres in good condition, plus spares 800 off. Engines only just run in, (with 300 spare engines thrown in); power assisted steering, disc brakes with brake 'chute for use in an emergency (plus 150 spare 'chutes); drivers seat (bucket type), ejectionable in an emergency, driver's seat area spacious with well laid out dashboard, speedo reads up to Mach. 2.5' passenger also fitted with ejectionable bucket seat; passenger position also equipped with the added luxury of radar, for use in fog.

Safety belt fitted in both seats, oxygen system installed for use when climbing high mountains also as a standard fitting (all models), intercommunications between driver and passenger, radio in good condition, V.H.F. and U.H.F., (will not pick up Radio Caroline); also fitted in this de-luxe model the very latest heating system with hot and cold air blowers.

Fitted with many extra's including - wheel trims, wing mirror, compass and fog lamp. 4 forward gears, but no reverse, also huge stock of servicing tools and equipment.

Crash recorder fitted for use at inquest and by Insurance Company.

Fuel consumption slightly above average - 1,000 gallons per mile, also fitted with overdrive (re-heat) when fuel comsumption rises slightly to - 1,500 gallons per mile, spacious fuel tank at rear end.

Cost new £ 500,000,000. will accept 30/-

Take advantage of this stupendous offer, it is not likely to be repeated.

S.A.E. in strictest confidence to:-

Mr. H. Wilson,
10 Downing St.
LONDON W.1.

Following the cancellation of the entire TSR-2 project, this notice appeared on the noticeboards at Warton...

Although humourous, one suspects that it sums up the feelings of the workforce.

development problems had peaked and therefore the government had very little to lose. All this good news appeared to have fallen on deaf ears as the government were still insisting on their figure of £604 million, (although later this did increase to £620 million). In July 1965, during a House of Commons debate on the subject, Denis Healey quoted a figure of around £750 million, though what was not made clear at the time was that this figure covered the cost of 150 aircraft, spares, training, and even the clothing for the aircrew to wear. Therefore this did not reflect the development costs alone as he appeared to imply.

### BAC Struggles Through.

The struggle by BAC to rebuild its crumbling empire after the cancellation and the disposal of the TSR-2 airframes is examined in the final chapter. There was however a whole series of seemingly endless problems and blunders to follow even after the announcement of the cancellation. One in particular involved the vast amounts of data that had been accumulated during the 24 flights of the single flying prototype. In the recent months before cancellation, the programme had been making up lost ground and much of the information that had been collected still required proccessing, and in addition a lot of cross checking still had to be done. The test aircraft carried two types of recording media, magnetic and paper tape and it was reputed that for each hour the aircraft was in the air, it required a whole day to interpret and analyse the results. In an effort to maintain the progress made, BAC made an approach to the government with a proposal to continue flying the two aircraft, XR219 and XR220.

Whether it was sheer lack of foresight, or more to the point, some other underlying reason, the government simply refused and Roy Jenkins, the Aviation Minister, announced in June 1965 that although he had considered the request by BAC, he could see no real benefits in allowing the two TSR-2 aircraft to continue flying at a cost to the tax payer of between £2 and £3 million. If however BAC were prepared to fund such a venture themselves, then the government would be prepared to let the aircraft be used for such purposes. The costs of such development flying would however be deducted from the forthcoming cancellation costs. This BAC could simply not afford to do. With its main military project gone every effort would be needed to ensure that its civil business would not also succumb to the ravages of cancellation. Already there were those wanting to review their investment in BAC, and unless the board could come up with a life saving plan, it was apparent that the company was doomed. Therefore spending £2 million on flying an aircraft that had no immediate viable or profitable future was, of course, totally out of the question.

Whether the government realised the opportunity it had in allowing the two aircraft to fly is clearly doubtful, but they could certainly have contributed a tremendous amount of valuable information to the industry, in particular to the Concorde project to which the government was now fully committed. It would also have helped those projects BAC were likely to embark upon in the future, either on its own or involving collaboration with other countries. At the time a rumour was circulating that the government were concerned with a possible detailed evaluation between the TSR-2 and the F-111 and that the shortcomings in the F-111 would be exposed. A special evaluation team was already in the States studying the F-111, and although the RAF were deeply involved with the TSR-2 as yet no service pilots had flown the aircraft. Had such a team have been allowed to fly XR219 there was just such a possibility a comparison could have been completed. The first RAF pilot to actually fly the F-111 did so in December 1965. The man was Wng Cmdr Geoffrey Fletcher and his flight had a number of problems with both engines flaming out when he was reducing speed from Mach 1.9.

**American Navy pulls out.**
By 1965 the United States Navy were becoming increasingly dissatisfied with the progress of their aircraft, the F-111-B, and there was talk of the Navy pulling out of the project. This of course put extra pressure on General Dynamics, especially with the accounting as the whole project was based on a production in excess of 1,700 units for American forces and a possible 1,000 aircraft for overseas customers. The F-111-A itself was also in deep trouble with the plane getting heavier by the day. The powerplant was giving problems, the leading edge lift device was unreliable and it was found there was a high drag problem at the tail end of the aircraft. All these problems, and more, had in fact already been highlighted by B.O. Heath, but as he was heavily involved in the design and development of the TSR-2, as the Design Project Manager, some may have felt that it was merely a way of discrediting the competition. It was obvious now though this was not the case, but it was too late. In Denis Healey's autobiography *'The Time of My Life'*, he mentions that he had to convince the Air Staff that the F-111 would suit their requirements far better than the TSR-2 and that they appeared to accept this, even saying they had wanted an American aircraft anyway! Whilst some may have changed their allegiance, many felt betrayed for they knew the F-111 could not fulfill the Operational Requirement. They had fought the Conservatives to get the TSR-2 off the ground. Now, according to Healey, they wanted American planes which were designed to fulfill a totally different operational role instead! When studying the specification of the F-111 there are a number of apparent differences in the roles the two aircraft were destined to perform. The F-111 had originally been designed as a 'Tactical Fighter', not a bomber. It also lacked many pieces of equipment that had been regarded as essential in the TSR-2, and the performance of the F-111 was well below that of the BAC aircraft. So, had Healey really needed to convince the Air Staff that the F-111 was the plane the RAF wanted, or was it in fact just a fait-accompli leaving the RAF no option but to accept the F-111, as there was to be no TSR-2?

It appears that the Chief of the Defence Staffs had outwitted the Air Force in the early stages of the TSR-2 programme and now the defence minister was doing the same at the end, with promises of a replacement aircraft that lacked the capability to carry out the roles in the operational requirement. What was concerning Healey, and had been since the February of that year, was the problems the F-111 was having. Although he knew these problems would eventually be rectified, he also knew that there would be a price to pay.

**Britain Cancels F-111.**
Toward the end of 1967 Denis Healey was again coming under increasing pressure to make more cuts in his defence spending. This time it was the F-111 that was on the line. The costs of the aircraft had now increased so much, that the bill for the 50 F-111s would at that stage, have been closer to the cost of 100 TSR-2s, and this did not take into consideration the modifications that would be required to convert the F-111-K to a British specification. Healey was in a quandary. On the one hand he had virtually forced the Royal Air Force to accept the cancellation of the TSR-2 by promising the F-111, on the other the Cabinet were once again forcing his hand to

## TSR-2 PHOENIX OR FOLLY?

cancel the American plane. Healey had even considered resigning on principle, but refused partly because he felt nothing would be gained by such action.

In the event Healey remained as Defence Minister and in January 1968 Britain informed America of its decision to cancel the whole batch of F-111s. This included a large amount of spares that had already been ordered and paid for. Subsequently these had to be sold back to the United States Air Force at a considerable loss to the British government. In order effectively to remove the requirement for an aircraft to fulfill the OR339/343 for good, the government announced its plans to withdraw from the Middle East and Persia even though there was still a considerable amount of unrest in those areas. There were also problems in other parts of the world including Vietnam and the Warsaw Pact countries where the situation appeared to be deteriorating. Britain's commitment to NATO would, perforce, continue to be carried out by the existing V-bombers.

The pattern of events after 10 years appeared to have come full circle, with the RAF still without a replacement for its now ageing Canberras. In the light of what had happened, it was decided to equip the Air Force with McDonnell Douglas F4 Phantoms and Mk2 Buccaneers. What Lord Mounbatten thought of this was anybody's guess, for it appeared the government were adopting the very policy he had recommended all those years ago. The Phantoms were to be fitted with the reheated Rolls-Royce Spey engine and a considerable number of British avionic systems. In 1964 the Ministry had been quoted just over £23 million to carry out such modifications to the Phantoms ordered for the Navy, but by May 1965, this figure had nearly doubled. The development of the engine alone increased from £12.5 million to nearly £29 million, with no apparent advantage. The Spey was heavier and much larger than the General Electrics J-79 units fitted to all American Phantom's as standard, and so the British versions had to be considerably modified at the rear end in order to accomodate the Speys. This not only pushed up the price and the weight of the Phantom, but also delayed the aircraft entering service with the RAF squadrons and the Navy.

*By 1978 the American Air Force were increasingly concerned about the amount of F-111s deployed in Europe and the lack of deep maintenance facilities. An exploratory contract was placed with British Aerospace at Filton, mainly for overhaul of the F-111 cockpit ejection capsule. The contract was renewed annually, but because the American's were extremely pleased with the quality of the work, in 1983 they offered British Aerospace a four year contract to carry out deep major overhauls and maintenance of the F-111s.*

# CHAPTER EIGHT
## People, Planes and Plans

Successive British governments, irrespective of political persuasion, have been responsible for the cancellation of many military projects over the years. The cost to the nation cannot always be measured simply in terms of money lost. In the majority of cases such cancellations came about merely to provide the government with a short term answer to a long term problem when trying to establish a stable fiscal policy. The many publications written on the TSR-2 have always indicated that it was cancelled because of the excessive costs of the project, and whilst this appears to be the most acceptable and logical answer, a number of inconsistencies appear which have led me firmly to believe that the 'cost excuse', although the most feasible, has been used merely to disguise an underlying, alternative reason.

### The Industry Under Criticism.

There has been a great deal of criticism aimed at our aviation industry, with particular emphasis on the large amounts of public money it was alleged to be using in order to sustain itself. The bulk of this criticism appears to have stemmed from the period just after the Second World War when the industry was trying to rebuild

*One of the two Bristol 188's undergoing construction at Filton. In the background English Electric T-4 Lightnings. There was nothing to show that the American method of aircraft manufacture was more efficient when dealing in such small numbers of aircraft? (Author)*

itself. This was the time of tremendous challenge with aviation pushing forward new frontiers, and research developing new technology.

The basic infrastructure of the industry had survived with aircraft manufacture following a long tradition, using a relatively simple method of building one or two prototypes to establish a basic design. This had a number of advantages. As the prototype was virtually hand built, its construction would expose any significant problems prior to the setting up of a production line. Secondly, an aircraft would be seen flying very early in the development stage, giving the investors the satisfaction of seeing something for their money.

On the military side, a large proportion of research and development work was being carried out under government instruction. The trend of building small development aircraft had been encouraged by the government in order to keep the various design offices busy, and also to retain the manpower and technology we had.

The main criticism though appears to have emanated from the controversy surrounding the Supermarine Swift and one must surely question the capabilities of the Ministry in its methods of procurement at the time. The project appears to have been overloaded with too many requirements and pressed into service before fulfilling an adequate development programme. The reformation of the industry had been brought about in an attempt to prevent a reoccurrence of the Swift episode and at the same time to streamline the workings of the industry in an effort to place more emphasis on work away from the governments control. At the time the TSR-2 design was reaching fruition, the reports being put out by the press appeared to be nothing more than speculative journalism. It is therefore hard to conceive that such speculation was instrumental in bringing about the demise of the TSR-2. Although when it came to the public's opinion of the bomber the media, of course had a great deal of influence.

Whilst the media criticised the industry and the amounts of public money it was receiving, rarely did one read of their successess and the millions of pounds it was earning the country in export orders. The aviation industry has consistently been amongst the leaders in overseas trade.

**Myth behind the Secrecy.**

There are many theories on the TSR-2 and the myth which has surrounded the project has led only to a great deal of speculation. If, or whenever, the official documents are released we may discover the truth. It has been strongly intimated in the Downey Report that these documents are quite likely to be reclassified for up to a further 40 or so years.

It was very often said that the TSR-2 was born at the wrong time and in the wrong country and that such a sequence of events would never have been tolerated in America. In many respects the latter part of this statement is true. The time was right for an aircraft of this calibre. However, it seems that the British governments had so little faith in its own aviation industry that, except possibly when faced with a full scale conflict, it was not sufficiently far-sighted enough in justifying, not only the cost of a project on the scale of the TSR-2, but also the costs of other similar imaginative projects in the period since the 2nd World War.

It has always been claimed that the sole reason for the programme being cancelled was one of cost. The subject itself is extremely complex and in the absence of any official figures, merely adds to the myth. Denis Healey's statement that the project costs were out of all proportion to its military value, surely had little

# PEOPLE, PLANES AND PLANS

argument, when at the time the world was in the middle of the Cold War and the expectation of war was very real.

The whole cost consideration has to take into account not only the development and construction of all airframes, but also the new high-tech equipment that was to be developed and produced alongside the airframes. The process does not stop here however. Once in service the support and maintenance costs throughout the aircraft's service life have also to be considered. In this regard British aircraft have had a somewhat poor reputation and previous such costs were rather high. This was something that was highlighted by the TSR-2 project. OR343 had stipulated the aircraft should be kept fully operational at a front line base with minimal support for a period of 30 days: a somewhat mammoth task for an aircraft as complex as TSR-2. Therefore many of its systems were duplicated or even triplicated. Thus it was calculated that the operational readiness of the aircraft would be superior to the F-111. The comparison between the TSR-2 and F-111, along with their differing costs, are studied later on.

### Was TSR-2 Justified?

TSR-2 has to be described as a pioneering aeroplane, and as such, any attempt to accurately predict its cost was virtually impossible. Although most of the costs had to be calculated in respect of a project that would come to fruition in a number of years time, the perception of those costs was inevitably based on current inflation and could therefore, only attempt to foresee both what was required and affordable.

In 1956 when a Canberra replacement was being considered, it was envisaged the new aircraft would take on the roles of the Canberra itself, and at the same time fulfill the new requirements, that so far were not being performed. This was to include a strategic defence capability, coupled with a reconnaissance role which could be carried out at low level with little regard to weather conditions. In order to

*It was obvious that Britain required an aircraft that was capable of maintianing low level flight in extremes of bad weather in order to reach a target deep within enemy territory. (BAe)*

do this would require a navigation and flight system that would ensure aircraft and crew safety. Also by including the equipment previously described in Chapter 3, would, because of the pin-point accuracy, ultimately reduce the number of missions required to achieve a successful outcome. At the same time this would also cut down on the numbers of aircraft needed by the RAF.

But was such an aircraft needed? At the time the Americans were building large aircraft able to fly at great height not only as a protection from ground attack, but also so as to be able to reconnoitre large areas of landscape.

In 1960, America had built special high altitude aircraft to carry out this role, but the U-2 incident soon changed the West's attitude. It was in May 1960 that Gary Powers on a reconnaissance mission over Russia, was piloting his U-2 spy plane at 60,000 ft. The Americans believed at the time, that Russia had nothing that could could touch them at that height. Powers was therefore taken completely by surprise when he was brought down by a missile. In fact during the 50's and 60's America lost 138 aircrew in spy planes flying high over Russia.

In 1963, NATO assigned the V-bombers to take on a low-level role. But these aircraft had never been designed to carry out such missions. Imposing such a role on them would have considerably shortened their service life through the fatiguing effects of low-level flying through the more turbulent air near the earth's surface. It was obvious then that the RAF now needed an aircraft that could penetrate, enemy territory with the ability to fly so low that the aircraft would go undetected by ground radar stations until the last possible moment.

There was talk of ordering two different types of aircraft, one to carry out conventional attacks on the battleground and another developed to concentrate on nuclear attacks deep within enemy territory. This whole plan had been suggested merely as an attempt to keep the costs down. It had been estimated that an aircraft intended to carry out successful attacks deep within enemy territory within the parameters of OR339/343, would require very sophisticated navigation and attack systems. To jeopardise such a costly aircraft by also allowing it to carry out a secondary support role on the battlefield was considered too expensive an option.

At the time the Air Staff appeared to be justifying a sophisticated bomber mainly because of the new missile technology. The versatility of the TSR-2 made it a necessary addition to the RAF's inventory and complimentary to the V-force.

**An Exportable Commodity.**

There appeared to be two areas of concern regarding the TSR-2, the main one being its cost and was it an exportable commodity which would allow us to reap back some of the development costs. The second, was there an alternative? The export potential of both our military and civil aircraft have brought the country much needed overseas funds. There is no doubt that the British industry has an excellent record of producing technology that has been favoured by overseas buyers, therefore there is no reason to assume that the TSR-2 would not have had attracted similar success. Although at the time it was the most advanced aircraft in the western world its export markets would be very limited. As the development of the TSR-2 progressed, so too grew the amount of interest, particularly from the United States. We have seen in the previous chapter the amount of interest shown by the Australian government. However, any possible sale resulting from this appears to have been thwarted by the British government who failed to grasp the significance of such a deal, and by a hostile Chief of the Defence Staffs, Lord Mountbatten.

# PEOPLE, PLANES AND PLANS

*The Hawk, one of Britain's recent successes in aviation exports. Still considered by many as one of the best trainers for high speed jets.
(Pete Holman)*

America, at the invitation of BAC, had made a study of the specification at a very early stage and had been both surprised and impressed with what they saw, even intimating that the United States Air Force could easily be interested in such a bomber. Like the Australians, the American government was kept informed of the progress being made. BAC's relationship with America had been nutured during the agreement to build the Canberra, (B-57).

The question of actually selling the TSR-2 to America did not come into it, as the American constitution forbids them to have any reliance on a foreign power for their arms. Further their method of procurement was completely different to that in this country. There was, however, the possibility of them building the aircraft under licence as they were doing with the Canberra. But, desirable as it was, this was highly unlikely, especially as they were still producing the B-57s. To import further high technology business in this form would no doubt, have been considered a retrograde step by those in power in America.

### An Ocean Apart.
The American arms industry is extremely complicated. Basically it is dependent on Congress which allocate funds for specific military programmes. Such programmes are not always necessarily based on a strategic requirement, but may be instigated because of a technological or even a political requirement. For instance, a Senator who may be seeking re-election in a State heavily dependent on military contracts, may be persuaded to campaign for a project the sole aim of which is merely to ensure that levels of employment are kept high. Another example is when a company may require government funds to develop a new concept in military strategy, and although nothing may ever come of the project, it

has at least ensured that military development continues unhindered and properly funded.

At the time, the American defence programme was coming under pressure. Up until this period, competition in this field of manufacturing had been limited mainly to industries within the United States, thus giving America a virtually unchallenged world market. This was to change significantly, with France and Russia in particular building huge arsenals of military hardware. Since then France has had a great deal of success with the many different variants of the popular Mirage and this together with the new breed of French missile, has made her one of the leading exporters of military equipment. Similar amounts of equipment being exported by Russia were quite alarming, with many Middle Eastern countries benefitting from the heavily subsidised weapons. So dependence on America for such technology was decreasing and this was causing some concern within the American government. This dependancy gave America certain manipulative powers, which, if the trend continued, obviously would lessen. This balance is studied later on.

The argument that American technology comes cheaper is never, I feel, a strong one, when the subsequent 20 plus years of dollar, technical and spares support is allowed for. As we shall see, the costs of the F-111 spiralled virtually in line with those of the TSR-2. The increasing complexity of military hardware meant that costs were certainly rising at an alarming rate. The many programmes instigated by America have already shown this to be true, with even more recently the Strategic Defence Initiative, (SDI) and the Stealth aircraft. Being dependent on a foreign country for defence systems would pose a number of problems and would make us very vulnerable in certain circumstances. By way of example I would like to quote from an article written in Air Pictorial in 1966, which clearly shows the manipulative power of America.

**American Influence.**

Lockheed supplied Indonesia with a number of the Hercules transport aircraft. When the Indonesian government needed to use its planes in anger during an emergency, the United States, presumably not agreeing with the course of action, blocked the sale of spares to Indonesia and in effect, grounded the whole Hercules fleet. Here surely was a case of the USA using its power and influence to manipulate another nation's affairs, merely because it had decided that the Indonesian government was acting in a manner contrary to the political views held by America. If America could do it to one country, then any other country reliant upon America for its arms, could fall victim to this same interference in its affairs. Military equipment does not always have to be involved, it may come in the form of ultimatums and the withdrawal of certain privileges.

It is a fact that even we appear to have fallen victim to America's whims on a number of occasions. The most noteable example of this is the Concorde. The failure of America to build a supersonic airliner resulted in the only project likely to succeed being the Anglo-French partnership which of course, had no American involvement whatsoever. The subsequent campaigns in America against Concorde gathered momentum until the airliner was banned from all major American airports at the very time when the Concorde sales campaigns were at their height. The American arguments at the time could not be substantiated as the TSR-2 had already shown there was little threat of damage to either health or buildings from the noise of the Olympus engines. Official sources in America appeared to do little in quelling the campaigns. As a result of this many of the world's major airlines were

# PEOPLE, PLANES AND PLANS

*The English and French Concorde prototypes at Filton. The campaigns in America against Concorde appears to have deterred major airlines from buying the worlds only successful supersonic airliner. (Rolls-Royce plc)*

deterred from adding the Concorde to their fleets. Therefore, the world-wide success of Concorde certainly seemed to depend on approval by America. Today, the two major airlines which began services using the supersonic airliner continue to do so and although they now use the major American airports, the aircraft has failed to attract the success its technology so richly deserved.

Such instances occur time and time again. When BAC attempted to sell refurbished Canberra aircraft, principally to South American countries, the United States government objected to the sale. The aircraft were surplus to an original NATO requirement and as such had initially been partly funded by the United States. Even so, this interference by America badly affected the whole sales deal.

### Technology Transfer?

Likewise, the costs of transferring this technological dependence has also to be considered. At the time, Britain's independence gave us control over the type of equipment we required and to a large extent the price we paid for it. Reducing the status of our manufacturers to that of satellite or remote job stations would give us very little control of the overall quality or type of equipment we could acquire. This would therefore limit us to the foreign technology that was available at the time and almost force us economically to adapt a vehicle to suit our needs, which was more or less what we were about to do with the F-111. Eventually of course the question of cost would no longer be a consideration, as our complete reliance on America for military equipment would be such that we would be unable to make comparisons in order to evaluate the true cost. Similarly neither could we continue to refuse the new technology because of its high costs. So doing would make us reliant upon equipment that would be outdated and ineffective against more modern systems. So, we would have either to switch our allegiance again or become totally dependent on America to place such weapons in this country that would ensure an adequate defence. The ratio of off-set trade, which so often had helped in the past, could no longer be used as a bargaining tool.

### Government Intervention.

When the government first intimated that it would award a contract, not to a single manufacturer, but to a consortium of companies, raised two separate questions. Firstly, there had been concern at the number of aircraft manufacturers and the amount it was costing the country to sustain such a large industry.

By forcing the separate manufacturers to form into two major manufacturing

# TSR-2 PHOENIX OR FOLLY?

establishments, British Aerospace and Hawker Siddeley, indicated that the government were prepared to sustain such institutions with substantial contracts. Eventually it was hoped that such reliance on government contracts would become less and less as each of the manufacturers established themselves in the world markets. The idea of a merger was not completely new as some of the companies had already sub-contracted work out to each other and had already considered such possibilities themselves. Now the government had come along to force through a process which would possibly have taken place anyway.

Secondly, the inducement factor of being offered the contract for the TSR-2 was, I feel, wrong and that the merger or the contract should have come first, preferably an amicable merger of companies. Although this would inevitably have delayed the in-service date of the TSR-2, it would also have allowed the manufacturers concerned time to create a unified structure and simultaneously build a firm foundation from which to design and construct the new bomber.

The three companies involved were used to a variety of working practices and industrial relations. The management teams were now having to coordinate the current day to day problems of the workforce and at the same time become involved in the complexities of the new merger and the new aircraft along with all their associated problems. On top of all this, the MoA, being the customer, began dictating certain conditions. The purchase and supply of specific components was to be taken over by the ministry. In doing this the effective control was taken out of the hands of the main contractor.

### Controlling the Contractors.

A multitude of contractors were employed on the project, (see Appendix 3), and whilst some were directly responsible to BAC, a large number had been commissioned directly by the MoA. These latter consisted mainly of contractors responsible for the Category 1 items which included the radar and reconnaissance

*The Harrier GR-1. With the formation of the two major manufacturers, Hawker Siddeley had hoped the P1154, which was based on the Harrier, would survive. The government again, failed to back the industry and the services.*

*(Rolls-Royce plc)*

equipment. Some of the Category 2 equipment, that approved by the Ministry but ordered by BAC, also came within this same scope. The engines, of course, were another problem as BAC had no say whatsoever regarding them, and they too had been ordered direct by the Ministry. There were problems being caused by the sub-contractors due to their dealing direct with the Ministry on such matters as cost overrun and modification approval. This non-standard arrangement effectively took away from BAC any of their own controls over such matters. Coupled to this were problems within the ministry itself. It is understood that around this period there were severe staff shortages at the ministry and very often an important decision would have to be taken by a junior member who had no real understanding of the problem involved mainly because of his lack of technical knowledge. Thus any resultant cost overrun or equipment modification was never accurately queried when, by rights, it should have been.

**Controlling the Committees.**
The 13 committees set up to control the project did not appear to be effective for very long, and in fact, some of the meetings they held had no representation from the manufacturer. So inefficient was the method adopted by the Ministry that in the end it was to prove quite costly, with all the blame and responsibility for such a policy devolving upon the main contractor. Coupled to this the research establishments which had been employed by the Ministry to assist with the project were attributing all their costs to the programme, when in actual fact TSR-2 was only responsible for a proportion. Many of the spin-offs that would help research departments understand problems encountered on other projects were also being billed exclusively to the TSR-2. The engine programme, fraught with problems as it was, was still making a massive contribution to the Anglo-French project, and although it was benefitting the French Concorde a great deal, the majority of engine development costs were also attributed to the TSR-2.

**Costing the Engine.**
The Olympus engines that were to power the TSR-2 were perhaps one of the most miscalculated aspects of the whole affair. When Vickers had first been awarded the contract as the prime contractor, they had wanted a Rolls-Royce engine. However, behind the scenes it was already understood that Bristol Siddeley Engines would supply the powerplants. This appeared to be because Bristol had conformed with the government's wishes and had amalgamated with Armstrong Siddeley. On paper, the idea of simply modifying an Olympus engine to fly supersonic appeared quite sound.

BSEL had submitted their estimate, at a figure of around £20 million. The ministry questioned this figure, especially as it was based on an engine already in production. It was suggested by Whitehall that BSEL should review their estimate. At the time the ministry were looking to achieve an engine price of below £10 million. Accordingly BSEL resubmitted their tender. It was considerably lower than their original figure and in fact at £7.5 million looked unrealistic. Then when the problems began to increase so did the price. What should have been a relatively straightforward project began to baffle the engine manufacturer, and the problems eventually increased the price to well over £32 million. Now whether Rolls-Royce could have developed their Medway engine any quicker and more cheaply is surely open to question. Quite naturally many of those I spoke to at Rolls-Royce Bristol were adamant that Derby could not have produced an engine with sufficient power within the same time period. Those I met at Derby tended to contradict their military colleagues at Bristol, insisting that given the opportunity the Medway would have been in the TSR-2 far earlier than the Bristol Olympus. But of course, this is purely academic now.

**Controlling the Build.**
BAC had appointed Henry Gardner as the Project Director, responsible for the airframe. Alongside him was George Henson, Project Engineer. These two men were to be responsible for aquiring the equipment necessary for the airframe, which was one of the biggest problem areas. Henry Gardner was a well known and respected man within the aviation industry, but the problems he was encountering had been inherited much earlier from a government whose views had been influenced by the Supermarine Swift affair and the Sandys' report. We have already seen how many very promising projects were cancelled because of this, and now the sub-contractors were becoming extremely wary of military contracts.

The anti-TSR-2 lobbying had already taken its toll and the large amount of adverse publicity was, not unnaturally, attracting the wrong responses. Being convinced that the TSR-2 would never make it into service, the sub-contractors began quoting higher than normal prices in an effort to ensure some return on what they considered would only be a 'one-off'. From this it should not be assumed that all the sub-contractors were being avaricious or irresponsible. Far from it. Because of the contractual procedures and the new working practices, problems were arising that hitherto had never been an issue and came about purely as an effect of the 'development batch procedure'.

*There are many similar examples outside of aviation where, in this high technology world, a low estimate has been submitted to secure a contract. Sometimes this is done in ignorance because of the complexity of new working methods. However, a close examination of the building industry will show that there is a great deal of optimism when submitting estimates which tend to increase considerably as the building progresses.*

**Development Batch Procedure.**
The MoA had insisted that the project should adhere to a production method introduced by Claude Gibb and later developed by Solly Zuckerman. This was based on the American aircraft production system, known as the 'Development Batch Procedure'. Although this method was introduced in Britain to increase the efficiency of aircraft production, it appeared to have had the opposite effect on the TSR-2 programme and it was therefore questionable that such a procedure could ever have worked to the same degree of efficiency in Britain as it was allegedly doing in the U.S.A.

Production runs in America were geared to much larger numbers of aircraft, with the F-111 alone expected to yield some 1,700 units. During the initial negotiations of the TSR-2 programme it was envisaged that 150 aircraft would be required, less than 10 percent of the F-111 run. Therefore with such a comparetively small number of aircraft involved, was the huge expenditure required in the setting up of such a complex production method really justified?

The French, whose aircraft production was somewhat larger than ours, continued to use the prototype system, as did the Swedes with SAAB. Britain had spent over four years and nearly £120 million on developing and building the aircraft before it first flew in 1964. Under the prototype system an aircraft could

# PEOPLE, PLANES AND PLANS

Ordered by the United States Navy the F-111B was to be built by the Grunman Aircraft company, a subsidary of General Dynamics. The aircraft was later cancelled because the American Navy lost all confidence in an aircraft whose development had so many serious problems. *(General Dynamics)*

have been flying perhaps 18 months earlier, incurring less cost and with the added bonus of being possibly much less conspicuous politically!

Building development aircraft in production rigs had a number of disadvantages. The first, of course was that there was a great deal of work involved before actually starting to build the aeroplane and this in turn delayed getting the aircraft into the air. The development batch procedure meant that all the complicated and detailed design work had to be completed before the rigs and tooling could be produced, and this quite possibly could lead to an over complex design in the airframe. Once the building had begun, work could then often be held up due to finding latent design problems in a component or, even with the aircraft complete, further inherent faults discovered during the flying phase could then mean expensive changes to an already committed production line.

It had been agreed that a certain number of components in the programme would be the subject of a fixed price agreement where it was considered an accurate cost estimate had been calculated. This, however, was not happening as changes in design meant delays and subsequent increases in the cost. Therefore it was agreed to implement an overall pricing incentive arrangement which included most of the main areas of development. Today, with the use of super computers, many of these design areas can be simulated and problems foreseen, taking much of the guesswork out of aircraft and engine design. One of the main problems in the 60's appeared to be with those components which had a long lead time. This meant that pressure was sometimes placed on the sub-contractor to work to a time scale that could quite easily change as the lead time parameters of other components around it changed. It began to appear as if the long lead time items were going to dictate the very progress of build, merely by the bottlenecks they were likely to cause.

It appeared therefore that the only real control BAC had over the project was in those areas of production for which it was directly responsible, but these amounted to little more than the airframe itself.

Was it really possible that the flexibility of the prototype system could have been replaced with the more rigid American method? Once the detailed design process was complete it left little room for inexpensive modifications once the tooling had started. There was no evidence to prove that the American method was either more efficient in its design and building methods, or that aircraft could be produced more economically, particularly when dealing in a production run of the numbers involved in the revised TSR-2 programme. The original development F-111, although outwardly looked like those in service today, bears no resemblance internally due to the continual modifications required in the programme.

**Weapons System Management.**
A number of publications have referred to the TSR-2's use of 'Weapons System Management', or WSM. This was a procedure, again imported from the United States, to ensure that the procedures and support intended for the TSR-2 during its development and build were totally compatible and not, as was usual in this country, cobbled up on an ad-hoc basis. However, it appears that any similarity between the American version of this procedure and that used on the TSR-2 was purely coincidental! An article in the October 1963 edition of *Flight International* said that, *'TSR-2 appears to have inherited all the bad points of the American system and none of the good'*. It seems that the ministry had altered and modified the procedure to such an extent that there appeared to be very little left of what had in the beginning, seemed to be an ideal system.

In America, such a procedure is used to advance the philosophies of defence technology, where possible using proven technology. Right from the beginning of a programme the prime contractor has full and complete responsibility for the whole project. A finely detailed contract is drawn up whereby the contractor will know exactly what is required of the project including all the equipment that will be needed. This will then allow him to accurately assess the project and enable him to agree an overall price.

As we have seen the ministry intervened so much that BAC was allowed to do very little on its own. *(Weapons System Management was used very successfully on the Polaris submarine building programme, following the procedure almost to the letter.)* This was the first occasion that a programme of this kind had been employed in this country. The procedure was continually reviewed and revised by the ministry but to such an extent that BAC had no clear and definitive plan whereby it could agree to an overall price. Indeed, the contract for the nine development aircraft was not signed until October 1960. Even then whatever price had been agreed was never going to be accurate enough as the project was undergoing continual design changes which were not settled until nearly two years later.

Had the scheme been allowed to follow the American tradition then BAC would have maintained overall control. Instead what happened was that BAC had control only over those contractors responsible directly to itself, and had very little

# PEOPLE, PLANES AND PLANS

effective control on those engaged by arrangement directly with the ministry. Although the ministry, in comparison, had very few contractors, it was most unfortunate that those they did have were 'key suppliers', responsible mainly for the specialised electronic equipment, including the radar systems and the manufacturer of the engine. Many of the MoD establishments were involved, mainly at the request of the aviation minister and these included the Royal Aircraft Establishment at Farnborough, the Royal Signals and Radar Establishment at Malvern and the National Gas Turbine Establishment at Pyestock. Here BAC had no control whatsoever over their expenditure within the programme.

Another factor which also delayed this process and added to the cost of the project was that at the time TSR-2 was being developed, there was very little technology or equipment available in Britain that was capable of being used in the programme. What we did have was incapable of operating at the high speeds and low levels demanded by the operational requirement. That which was available was heavy and cumbersome and would have resulted in a larger and much heavier aircraft.

Many of those working on the project had little faith in the industry surviving beyond TSR-2. To make matters worse, the many committees formed to control the project costs appeared to have failed in their aims for a number of reasons. Apart from Concorde, TSR-2 looked to be the last significant aviation product for some time, and accordingly was attracting anyone and everyone who had anything to do with aviation. The repercussions of this were that very often the many meetings were over attended with the result that the true purpose of the meeting was frequently overshadowed by frivolous discussions on a minor technical matter, which in any case could quite easily have been rectified in a few minutes by any competent engineer.

*Weapons System Management should have ensured TSR-2 did not encounter problems from the past. Unfortunately because of so much intervention by the Ministry WSM failed to achieve its goal. (BAe)*

# TSR-2 PHOENIX OR FOLLY?

**Forcing the Issue.**

What of the government's attitude to the TSR-2? The urgency on the part of the Conservatives in forcing the manufacturers to merge appears to have been their main preoccupation, not necessarily the aeroplane itself. The Air Staff had put forward a convincing argument that the country needed an aircraft as defined within the parameters of the Operational Requirement. Although Peter Thorneycroft appeared to accept this, he was never fully convinced that such an aircraft would be a viable proposition. Again, in 1962, the government had been assured that the TSR-2 was the most important military project and that they must ensure it was built. Even so, there appears to be some foundation in the theory that a Conservative government, if re-elected, would have cancelled the programme. Indeed, in 1963, the Observer newspaper were already advocating this and had put out statements to that effect. By this time it appears the project had achieved its main objectives, namely - reducing the number of individual manufacturers, and TSR-2 on the face of it, appeared to be nothing more than an expendable inducement on the way. But perversely the programme had now seemingly got out of hand with the TSR-2, having apparently met all the conditions laid out in the operational requirement! The Conservatives were also being politically embarrassed by a number of indiscretions within the party, and to consider cancelling the project at this point would have led to even further ridicule in the light of the forthcoming election.

The Socialists, on the other hand, were becoming concerned at both the costs and the roles being mapped out for the new aircraft. The main concern being voiced by the Left Wing factions of the Labour Party, was that the TSR-2 would be used in conjunction with nuclear weapons. It was the age of 'Ban-the-Bomb' and any association with things military, nuclear in particular, was looked on as taboo. With no access as to what was really being spent on the TSR-2, the majority of Labour's arguments at the time could only have been based on hearsay and what was printed by a poorly informed press. The figures being bandied around at the time were anything between £240 and £260 million. In fact, little more than £125 million had been spent on research and development at that stage, against an original estimate of £90 million. Even had they known the real costs, there is little doubt that the Socialists would have continued to condemn the TSR-2. The bitterness and preoccupation of the Labour Party in condemning it as a 'Conservative Prestige Project' appears to have been far more overwhelming than any desire to see Britain with an effective defence system. The bitterness in the Party toward the TSR-2 prior to their victory in 1964, was already well known. The Shadow Aviation Minister Fred Mulley, had stated at the time, that he had no intention of retaining the post once the Labour Party came to power. He was kept to his word and actually took the post of Minister of Energy, leaving a reluctant Roy Jenkins to look after aviation.

**The Navy Misses Out.**

There were other defence projects underway at the time which appeared to escape the severe cutbacks facing the aviation industry. The contract for the Polaris weapons system had been hurriedly signed in 1962 after the government had failed to secure an amicable agreement on the cancelled U.S Skybolt programme. Polaris was not coming cheap as it required additional funding by the government. The system needed a platform from where the missiles could be launched, and as Britain at the time had no suitable vessel that could deploy such a weapon, the government had to embark on a costly building programme to construct five nuclear powered submarines.

It was alleged that Polaris was politically advantageous and that the forthcoming build programme would help the beleagured shipbuilding industry. This view seems to have been upheld by the Labour government when they made no mention of cancelling the American system during their early reviews when seeking to reduce defence spending. Why ?

Polaris, being a wholly nuclear defence system, quite blatantly contradicted the traditional Labour thinking on defence policy. However, a decision to have cancelled Polaris, while at least ensuring adequate funds for the TSR-2 programme, would have had far more serious consequences. No doubt Britain would have faced heavy cancellation penalties, but there was also the real possibility that there would be far more reaching repercussions.

Although Lord Mountbatten was confident he would get Polaris, he was also aware that the basic cost of the system and its associated expenditure, along with the forthcoming carrier replacement programme, and the cancellation of TSR-2, would cause a serious rift between the air force and the navy. Furthermore, it would also cast serious doubts on the integrity and ability of the Chief of the Defence Staffs to remain totally impartial.

TSR-2 was a front line deterrent and would be used in the initial stages of any conflict, conventional or nuclear, unlike Polaris which was designed as a second line of defence in a purely nuclear theatre of war. The general opinion was that the combination of TSR-2's multi-role flexibility along with the proposed through deck carriers, together with the current military equipment would provide Britain, not only with an adequate defence system, but at the same time also enable her to perform her role in NATO.

**A Change of Command.**
Having taken office in October 1964, the bulk of the Labour government's problems appeared to be monetary. The state of the accounts was worse than they had anticipated and unless the pressure on the pound could be alleviated, the government would be forced to devalue. Such measures would of course have a derogatory effect on the government's proposed spending and on all current projects. Therefore, the planned visit to America in the December by a British delegation was to have a profound effect, especially upon Britain's aviation industry.

Perhaps the most frustrating thing was the Labour government's irrational attitude in dealing with the TSR-2 when it came to power in 1964. While we have to accept that to some degree those in responsible positions outside of the government, particularly the Air Council, had adversely influenced those new members directly concerned with the TSR-2, there were certain instances when the complete story was never objectively put forward. This alone cannot excuse those government representatives who had promised an exhaustive independent review of the project which would involve the main parties concerned. Whilst using methods conceived in America to build the aeroplane, the idea of using the stringent forms of independent review, as used in America, before cancellation was clearly not to be.

The Plowden Committee had been organised purely to study the aviation industry as a whole and not individual projects such as the TSR-2. Whatever the reasons for this committee it failed to bring about any changes that had been

*When Wilson and his Labour Government came to power in October 1964, the industry were poised for further radical changes because of Labour's irrational attitude to the aircraft industry. (Authors collection)*

expected within the industry. The committee had no power to either make changes or instigate new proposals and therefore certainly gave the impression that they had been set up purely to vindicate a poor Labour policy on defence and aircraft manufacture.

Although the committee highlighted certain advantages in the TSR-2 over the F-111 regarding its cost and support, it appears that the government chose not to listen. Cabinet members spent many hours at Chequers discussing the various cancellations, but it appears that at no time were the industry consulted where their presence could have resulted in an objective case being put forward. In this way the government failed to reach a fair and proper conclusion.

BAC's awareness of the government's thinking regarding the TSR-2 and of the consequent vulnerability of the project had prompted them into making a number of approaches to the government in an effort to make it aware not only of the need for an aircraft of TSR-2's capabilities, but of the possible consequences to the industry as a whole should TSR-2 be cancelled.

BAC was still a relatively new organisation and the TSR-2 had been the catalyst in its formation. Should this contract now be lost, the company would surely suffer from an ensuing lack of confidence both in this country and abroad which, ultimately, could affect the future prospects of both home and overseas orders. If the British government was not prepared to show confidence in the company, what chance had BAC got of convincing overseas markets to invest in its products? Already there were those in the City who wanted their investment moved elsewhere. When the strong rumours of cancellation were being spread around, BAC had been urged to act quickly and put forward for consideration a fixed price proposal to the government for a specific number of aircraft, and even being prepared to lose some £9 million themselves by doing so.

We already know that Wilson's position was being compromised by the debts inherited from the previous government and that he was fighting desperately to avoid devaluing the pound. Any assurances that the cancellation of the TSR-2 would somehow immediately help the government's financial situation could only have come from those parties who had more interest in the demise of TSR-2 rather than its success. Having used the TSR-2 as a tool to their advantage during the election campaign, Labour had to continue the pressure and now be seen formally to condemn what they had always stated they considered to be nothing more than Conservative 'dreams of grandeur'.

**Pressure from America**
The visit to the States by the British delegation in December 1964 was, in my opinion, to have a significant effect on the outcome of the TSR-2!

What was negotiated in America can only be a matter of conjecture, however there is increasing speculation that the success of Britain's application to the IMF all but depended on the cancellation of the two major aircraft programmes in the UK, namely Concorde and the TSR-2. For some time America had been trying to manipulate Britain into a position whereby we would be dependant on America for our aviation needs, including military aircraft, and this now appeared to be their opportunity.

Since the 2nd World War America had watched the development and growth of military aviation throughout the world. The allied countries were now independantly investing a great deal in an effort to replace and refurbish their war losses. This was causing America some concern insofar as they had been the main source of this equipment during the war. This trend could adversley affect demand and have a destabilising influence on the economy of the USA and the growth of their military establishments. America alone could not sustain such a market even though in that same decade diversification had come in the form of the space race. Whilst the space programme would use limited sectors of the industry, the demand on the financial resources was out of all proportion to its size.

In order to resolve the matter America, instigated a huge sales campaign in Europe to sell its military hardware. Henry Kuss, a unique and powerful salesman in America, had been given the task by Robert McNamara. It appears the arguments used by Kuss were based on the theory that the size and resources of the American aviation industry were so vast that this alone could eliminate any competition from Europe. Therefore in order to reduce expenditure on military R&D and manufacture, it was anticipated that Europe would resist the struggle to compete and simply purchase their requirements from America. Participating countries would be given the opportunity of becoming involved by having a limited share in any such building programme. Britain had been one of those specifically targeted by the campaign, but so far we had strongly resisted. However, in 1964 Sterling was under pressure. We could devalue but that would have shaken confidence in what was at the time the world's reserve currency. So Britain turned to its ally, America, to help support Sterling and this in my opinion, was a major factor in the demise of TSR-2.

America's dealings with Britain had in the past, met with with a great deal of resistance. The Conservatives had severely opposed any pressure from America in its negotiations, specifically in matters dealing with aviation. It was with this inflexibility in mind that America prepared to negotiate with the Labour led delegation. However, publications that refer to this meeting indicate that such opposition never materialised and that America experienced very little resistance to their demands. There were allegations that Harold Wilson lacked the business

*Robert McNamara seated with the American President Lyndon.B.Johnson. Behind them are the Defence Chiefs. (Yoichi.R.Okamoto - LBJ Library)*

experience and negotiating tactics of those who had been before him, making him vulnerable during the negotiations in December 1964. America were determined to see the downfall of Britain's aviation industry, but equally so was Wilson's own determination to ensure that the pound was not devalued. These factors together appear to indicate that the TSR-2 had indeed been sacrificed merely to provide an agreement.

The actions of the Government on its return from the U.S.A add to this theory. Whilst Wilson insisted funds should be made available for the programme, there were those that opposed this action, indicating that they perhaps had knowledge of the pact and therefore did not want to see money wasted on a project that was on the verge of cancellation. However the charade had to be maintained until such a time when the cancellation process would not be adjudged to be part of any conditions imposed on the Government during their talks with America. Having previously laid the foundations for cancellation in their election campaign, an announcement in the Budget Speech appeared to be an ideal opportunity.

As an extra inducement Britain would be offered the F-111 at a special price. The various campaigns in Britain against TSR-2 would have encouraged Wilson to support such a demand. Allied to this were his own views on the industry which did not appear to be particularly supportive, and very conveniently the purchase would fulfill McNamara's wishes for a highly reputable customer for the F-111.

It is my opinion that the British Government, having succumbed to American pressure, had to maintain that the TSR-2 was being cancelled because of its increasing costs. The consequences of revealing the truth to the nation would have had a catastrophic effect on the newly elected Government. Such an admission would have forced a motion of 'no confidence' in a Government who were seen to be bowing to foreign intervention, and therefore such action would not be tolerated. The deal concocted by the United States government whilst enabling Britain to obtain funds from the IMF, had theoretically enforced the elimination of the competition for the F-111, but more importantly, it had fulfilled America's other ambition, the decimation of the British Aircraft industry. This actively encouraged some of Britain's most experienced and skilled designers and engineers to seek refuge with the American aviation industry. No longer, it appeared, would America have to consider Britain as a leader in aviation design and manufacturing for the simple reason that they had now made available some of the world's finest people to choose from.

The loss to Britain went far beyond the cost factor. The previously announced cancellations of the HS681 and the P1154 had already put a great strain on the industry as a whole. The demise of the HS681 alone meant the eventual closure of the Armstrong Whitworth works in Coventry. Almost immediately in excess of 12,000 jobs vanished with many more following over the ensuing months.

A new avionics laboratory had been set up by BAC, which was not only looking into equipment for the TSR-2, but was also carrying out research work intended for future programmes, also became a victim. The BAC works at Luton were eventually to close with the work being moved to various sites throughout the country. With little work expected within their specialised fields, the outlook appeared extremely gloomy for many of the smaller firms which were left facing financial ruin. Most of the people affected could no longer find employment in their particular field in this country and therefore faced either having to find employment using entirely different skills or even perhaps going abroad. Engineering, electronics and many other workers whose high-tech skills would be appreciated far more overseas were soon heading toward America, which welcomed them with open arms. Such were the consequences of leaving Britain drained of those resources, that when and if the time came for her to rebuild her crumbling aircraft industry the men to do it would be gone and the learning process would have to begin all over again.

**Arguments in Parliament.**
The ensuing arguments and discussions did very little credit to either the government or the country. On April 13th the debate on the Budget in the House of Commons was nothing short of pandemonium, with the discussions on the TSR-2 receiving most attention. The Secretary of State for Defence was still refusing to answer the question of commitment on the F-111, saying only that America had offered a firmer price than had BAC. What he meant by this is unsure as the cost of the F-111 itself was continuing to rise as its problems became more and more complicated. Had the government deluded themselves or were they trying to play down the fact that the price of the F-111 no longer represented the huge savings it had promised when they had negotiated the deal in December? It had already been stated that buying aircraft on a cost plus basis was not only too expensive, but also very unpredictable, and yet here it was ordering an aircraft that, as early as January, was undergoing severe development problems. The Government were basically doing exactly what they had just told the country it could not afford to do with its own industry!

There also appears to be some confusion as to whether the government had in fact budgeted for any unexpected price increases in the cost of the F-111. For some reason they were not prepared to divulge specific information on the F-111 contract, revealing only what they felt would appease the opposition and the public. It is almost certain that the savings expected by the cancellation of TSR-2 were already being taken up in the increased cost of the American aircraft! If this is true, then the government really had made an inexcusable error of judgement in its calculations for the replacement aircraft and the British industry was now paying the price of shoring up the pound. Ironically a year later the government lost its fight and was forced to devalue the pound. Again, this would make any real comparison costwise between the two aircraft extremely hard to calculate.

For some reason a gag was to be put on many of those who wished to speak favourably on the merits of the TSR-2. In particular Roland Beamont, who had been invited to a lecture for the Royal Aeronautical Society prior to the project being cancelled. Because of the nature of the subject, Beamont was requested in the normal way, to submit his paper for security clearance. I have listened to this gentleman on a number of occasions and I have found his talks extremely interesting, being factual and to the point. Under the circumstances he himself would have omitted any material likely to breach the security regulations at the time. Even so, the talk was inexplicably cancelled by the RAeS, apparently as the subject was being illustrated as the ideal vehicle for the RAF, which was contrary to the government's plans. Denis Healey on the other hand, appeared free to make statements that would be uncontested. Through selective quotations the TSR-2 could be made out to be as vulnerable as any other aircraft. By focusing on the new anti-aircraft weapons it was hoped to demonstrate how the TSR-2 would be open to a successful attack. Even more remarkable were comments made on the weapons TSR-2 was to carry, declaring that it would only be able to carry ordinary high explosives capable only of blowing up bridges and tanks. Whether this last statement was to appease the Left Wing factions or not, it typifies the irresponsible attitude taken by many within the party. Throughout the arguments for and against TSR-2, statements were made that in all honesty, simply could not be substantiated especially by such an apparently ill informed government.

### Questioning the F-111.

Putting cost aside for the moment, was the F-111 really the vehicle to replace the TSR-2? An article appeared which examined the Labour government claims that the F-111 was in fact the superior aeroplane and that this was the real reason why Britain so readily agreed to its purchase. Although the author of this article was obviously pro Labour, this does not excuse his failure to carry out basic research

There is increasing speculation that the F-111 would have failed to meet the requirements of the RAF in more than one area. Likewise the true cost of the aircraft appears to have been deliberately hidden. *(General Dynamics)*

# PEOPLE, PLANES AND PLANS

into the differences between the two planes. Had he done so, he would surely have seen that the roles mapped out for the TSR-2 could not be effectively carried out by the F-111 and therefore the question of either aircraft being superior was not a valid one. Although in this case the accusation had been intentional, there were those confused by the issue especially as there was an element of overlap in the roles intended for both aircraft. The requirement for the F-111 had been issued in 1960 under a guise very similar to the operational requirement issued for the TSR-2. However, the F-111 was also required to fulfill the needs of the navy with a Fleet Air Defence variant.

The actual build programme for the F-111 got under way around the same time as the TSR-2, with the aircraft first taking to the air in the December of 1964. However, the F-111 development was perhaps not as intensive as that of the TSR-2. The American aircraft was built using well proven methods practised by the industry over a number of years. Also most of the components that went into it were already in production. The engine was a tried and tested unit, already powering military aircraft, and only requiring an increase in thrust rating to achieve its goal. Most of the avionics, with the exception of the terrain following radar, were again basically off the shelf modules that were adapted around the needs of the F-111. The component requiring by far the most development, was of course, the variable geometry system, which had never been used on a production aircraft.

Apart from the obvious differences between the F-111 and the TSR-2, the biggest changes were in the on-board systems carried by the two aircraft. The F-111 had a terrain-following unit that included a comprehensive multi-mode function, it also carried a nav/attack subsystem with an inertial navigation system. It did however, lack many of the essential components that made the TSR-2 unique. Later different variants of the F-111 were to include some very sophisticated equipment that enhanced the aircraft's roles considerably. In most cases this meant a different aircraft performing a specific role, something denied to the RAF.

The capabilities of the F-111 are also questionable when compared to the performances insisted on in the Operational Requirement issued by the Air Ministry. There are no figures readily available for the American aircraft. However, the consensus of opinion was that in times of real hostilities the mission parameters would be designed to suit the aircraft and its payload. This seems to indicate a limitation in its range restricting its capability. The internal tankage of TSR-2, together with a ventral and wing tanks, would allow the aircraft to strike a

A unique feature of the F-111 was the swing-wing. *(General Dynamics)*

# TSR-2 PHOENIX OR FOLLY?

target some 1,480 nautical miles from base at Mach 0.9. By using in-flight refuelling this range could be extended by approximately 650 nautical miles.

The question of speed also appears to have been sidetracked. Although most versions of the F-111 can briefly exceed Mach 2, this can only be achieved over 30,000ft, using full reheat and in a clean configuration, when the skin temperature monitor exceeds the number of seconds allowed then the speed is reduced before any damage is incurred to the airframe. It is argued that such speeds have no real value as the benefits are far outweighed by the fuel consumption. TSR-2 had been designed to carry out a high level sortie at Mach 2 with a target range of some 550 nautical miles.

These then are some of the more significant differences between the two aircraft, but I feel that the biggest advantage of the F-111 over TSR-2 was adequately summed up by Roland Beamont when he said, *'the F-111 was not cancelled'*.

### An Element of Doubt.

A question mark continues to hang over the information which was given to the British government regarding the specifications for the F-111. There is a suspicion that the Americans may have purposefully misled the British delegation in their efforts to sell the ailing bomber, merely to secure a major overseas customer at an early stage. We have already seen that the equipment content of the American aircraft was not as comprehensive as that in the TSR-2 and one must therefore doubt the ability of the F-111 to conform with the operational requirement. America had always been conscious of the capabilities of TSR-2, having been kept informed by BAC. With this information available America had to ensure that the F-111 was

*What might have been! This illustration depicts an F-111K converted to a British specification. Like the imported F-4 Phantoms, the F-111 would have required extensive modifications in order to make it acceptable for RAF service. The modified in-flight refuelling probe is just one of the many changes that would probably have been necessary. (Phil Kingham)*

seen to be at least as good as, if not better than the TSR-2. One particular factor which has to be considered is America's record in the art of selling. If there is one thing at which America excels it is salesmanship and one government that appeared to succumb to their tempting offers was the West Germans. The West German government purchased a large number of the notorious F-104 Starfighter at a time, when, it now freely admits, there was no specific role for the aircraft to play! The Starfighter's reputation quickly gave it the nickname, 'Widowmaker', because of the number of pilots killed in accidents in which it was involved.

Britain's situation appeared to be similar to that of the Germans. She required aid and America needed to sell its products, therefore America was in a strong position whereby it could induce Britain to purchase goods it did not necessarily need or want. In fact Britain may have had little resistance in agreeing to a sale based on, what appears to have been, misleading information regarding the specification, performance and cost of the F-111.

The aircraft designated to the U.K was the F-111K, which in fact was never built. Those in the schedule were reallocated and became F-111As. There have been no precise details as to what modifications would be required to adapt the F-111K to RAF standards, therefore we must assume it would be items such as radio, TACAN and perhaps other significant changes in instrumentation and in-flight refuelling facilities. The fiasco that beset the conversion of the F-4 Phantom for RAF service gives us some indication of what could also have befallen the F-111 and which would certainly have damaged its merits even further. There had been some controversy over the number of way-points required by the RAF for the central computer in the TSR-2. It had been strongly suggested that, in conjunction with other navigational aids, the number of way points was somewhat of an overkill, and that the system would be equally as effective using half the number. However, the RAF were undeterred and continued to insist on the specified number. Ironically the F-111 did not even have such a facility and, thus, the RAF would have had an aircraft that lacked the facility altogether!

**Blackburn's Attempt.**
Did Britain need to go abroad for a replacement aircraft? Roy Jenkins and George Brown appear to be the only Cabinet Ministers who asked that the Cabinet consider a British alternative. Apart from that there seems to have been no design other than the F-111 discussed, or even put forward for consideration. In not doing so, they effectively eliminated any British aircraft that perhaps could have made a suitable contender.

Blackburn, in line with Government policy, had amalgamated with one of the new major organisations, Hawker Siddeley. The Blackburn company itself had continued to carry out extensive development and research using the Buccaneer airframe. Originally the Buccaneer had been designed as a subsonic aircraft specifically for the Royal Navy. The Navy had dismissed the idea of going supersonic, mainly because such an aircraft would be heavier, due to the extra equipment necessary, plus the need to carry extra fuel. It was a fact that this would only have resulted in causing extra problems with getting the aircraft off a carrier deck. It did, however, have a rugged airframe, which was able to withstand the severe buffeting expected during low level flight. Blackburns had also put in a tremendous amount of effort to ensure the aircraft would be comfortable to fly for the crew, and had even succeeded in eliminating the natural frequency levels normally associated with air sickness at low altitude.

# TSR-2 PHOENIX OR FOLLY?

*The Blackburn Buccaneer. It appears Blackburns attempt to compete with the TSR-2 was never going to be technically feasible. (Rolls-Royce plc)*

A series of modifications had been drawn up in an attempt to capture the OR339/343 contract. These included a thin wing, reheated engines and a completely new nav/attack system, with a head-up-display. Modifications to the undercarriage had also been drawn up giving the Buccaneer a twin bogie arrangement in order to comply with the rough field landing requirement. In fact very few considerations appeared to have been overlooked in the company's attempt to satisfy the Operational Requirement, particularly with regard to the proposed power plants. A supersonic version was already being considered for the RAF, where the Buccaneer would be equipped with an engine and reheat system very similar to that fitted in the British version of the F4 Phantom. The basic airframe had already been proved, as had the engine in the Phantom. The Buccaneer used lift augmentation, having lift aids across the leading edge of the wing and also in the tailplane. Ferranti at the time were using a Buccaneer as a flying test-bed to prove various avionic systems that were to be fitted in TSR-2. Blackburns themselves, were convinced that what they had appeared to be an almost ready-made replacement for the TSR-2: an aircraft that was well beyond its basic development stage and needing only limited funds in order to take it to the next phase of development.

Although this was an admirable attempt by Blackburns to compete with the BAC aircraft, it seemed that what was actually put forward appeared to be little more than an optimistic sales campaign and was never going to be technically

# PEOPLE, PLANES AND PLANS

The McDonell F-4 Phantom. Purchased by the British government, firstly for the Navy and then to help fill the gap left by the cancelled TSR-2. The British version of the F-4 was fitted with the Rolls-Royce reheated Spey. A superb shot of an F-4K from 74 squadron shortly before the aircraft type was withdrawn from RAF service. *(Motordrive Photographic)*

feasible. So many radical design changes would have been required to give it both the necessary speed and range that ultimately it would have resulted in a Buccaneer that bore little resemblance to the original submission.

## Was it the Cost?

These, then, are some of the key issues that dogged the programme throughout its rather short life, and leads to the inevitable question of costs. This is perhaps one of the most confusing and controversial issues of the whole story, especially as yet again there are no official figures available to substantiate any theory. Also, there are a number of considerations that have to be taken into account when dealing with this subject. The cost of development and the inflationary effects on the programme, are perhaps the ones we are more familar with, but also to be considered was the fact that, at the time there appeared to be a cheaper replacement in the F-111. Likewise, the cost of importing the American aircraft and the question of off-set sales must also be taken into account. We also have to consider that up until 1967 the pound was over valued and the devaluation would have affected the proposed off-set sales. One aspect which is rarely considered is the invisible earnings of the Treasury through direct and indirect taxation.

If we look at the cost consideration first of all, forgetting the element of competition with the foreign aircraft, we can see the government were originally talking in terms of 150 aircraft. It was calculated that each would have a unit cost of £5 million and that this would include the supply of all ancillary equipment.

Because the MoD was forced to review its spending, the number of aircraft was renegotiated and a total number of 100 units was then to be considered. (These figures do not include the 9 prototype aircraft and would therefore have been 159 and 109 respectively).

The non-recurring costs such as airframe research and development on the nine development aircraft were, as we have seen, to total some £210 million. The research and development costs of the equipment ordered by the ministry covering electronics and engine were approximately £60 million. This gives a total launch figure of something around £270 million. The production run based on 100 units at £2.1 million per airframe plus £1.3 million for engine and electronics, now gives us a grand total of £610 million, (£270 million launch costs, £340 million for production). By March 1965 this figure had increased to £650 million, but after this there is very little information to support further increases. Had we gone ahead with the TSR-2 Denis Healey claimed that the cost of the programme over a 12 year period would have cost the nation in excess of £1200 million. There is no comparable figure available for the F-111, but it does not seem unreasonable to assume that such costs would be no less after taking into consideration reliability, the huge expense of importing spares and dollar purchase.

**Spiralling Prices in America.**
Let us now consider the implications of the F-111 purchase. As stated, the deal was for 50 aircraft only as it was intended to support the F-111 with the Anglo-French Variable Geometry aircraft or more F-4 Phantoms. In December 1964, it had been estimated that the 50 F-111s would cost approximately £125 million, a unit cost of £2.5 million, I agree that on the face of it it does appear to be a much cheaper weapon if, as we had been led to believe, it could perform the duties required. To assist Britain there would be an off-set sales agreement whereby America would, directly or indirectly through a third party, purchase British goods.

There was some debate as to whether this form of aid would actually work or even, in fact, if it was at all valid, as it was based around the agreements timing. Some were concerned that there would be large time-scale differences over the sale and theoretically this would give no real benefit to the off-set trade deal. Likewise, those deals with third parties were also questionable as it was felt that such transactions would have been conducted anyway, without American intervention. The calibre of goods also has to be a consideration and again it was felt that unless we traded in commodities of a similar value there would be little impact on the off-set trade agreement. At the time the ability of Britain to export goods such as aero engines to America was virtually nil as their market was adequately provided for by its own manufacturers. So the question of off-set trade appears to have been little more than a token gesture.

There were rumours circulating at the time of the Saudi Arabian sales campaign in 1964/65 that America had purposefully withdrawn from the competition in order that Britain could secure the contract more or less unopposed. In this case such a gesture would still leave the competition open to Britain and France, but an influential word from Amercia in the right Saudi Arabian quarters would help secure the contract for Britain. The sale of military equipment would certainly give Britain the opportunity to earn much-needed dollars which in turn would help to pay for the recently ordered American F-111s.

Denis Healey was aware as early as February 1965, that the F-111 was undergoing severe difficulties. A number of government representatives had been despatched to Fort Worth in an attempt to assess the programme. What they found was a project confronted with so many problems that, at the time, they appeared unsurmountable, and the subsequent report to the government cast serious doubt over the aircraft itself. Ironically, shortly after the visit, America requested assistance from Britain. Dr Seddon from the RAE Farnborough was sent out to General Dynamics to help overcome some of the more serious defects, in particular, the engine stall problem. These setbacks were to effect dramatically the overall cost of the aeroplane. When Harold Wilson was revealing some of his forthcoming plans in early 1965, the costs had risen considerably. In April when the cancellation of the TSR-2 was announced, it was estimated that the F-111 was already costing around £5.5 million per aircraft. Although the government had produced figures to show just how much would be saved by the cancellation of TSR-2, this no longer appeared to be the case. It was calculated by 1968 that the price of the F-111 had nearly trebled to in excess of £425 million and was still rising. This figure is based on the government's announced estimates on the savings that would be made by the cancellation of the American aircraft, which was £425 million over a ten year period had we gone ahead with the project, and appears to have been a gross miscalculation. It can be seen how the price had got so much out of hand, but it did not stop there.

**Talking to Europe.**
I mentioned earlier the intention of the government to supplement the F-111 with a number of the proposed AFVG. There is little to suggest what the cost of this aeroplane would have been and therefore it is extremely difficult to assess what increases would have been added to the defence budget. However, as the AFVG was to be a new aircraft, it is not unreasonable to assume its unit cost would be relatively high even though it was to be built in conjunction with the French. This then makes even 150 TSR-2 bombers look considerably more economical overall. The cost of cancelling the AFVG was approximately £2.5 million and this must, I feel, be included in all the calculations within the defence spending for that period.

In 1968 the British government announced to the United States government that it no longer wished to proceed with the purchase of the 50 F-111 aircraft. The government had however already purchased a large number of spares and had to sell them to the United States Air Force at a considerable loss. Overall, the cancellation of the F-111 was to cost the British taxpayer over £46 million, and if we add to this the AFVG cancellation costs it makes nearly £49 million. The cancellation costs of the TSR-2 were still being calculated, but at the time stood at around £195 million and were still rising. Therefore, over a 3 year period the Labour government had wasted over £244 million of British taxpayers money and all we finally had to show for it were two airframes in museums!

**Hidden Advantages.**
We now come to the odd variables in the complex equations. Even if the F-111 deal had gone ahead, there was still the cancellation of the TSR-2 to take into account. Earlier, I based my figures on around £8.5 million for each F-111, so the 50 overall would be £425 million. Add to this the TSR-2 cancellation fee, which at the time was £195 million, and we arrive at a figure of £620 million, which if you remember was about the price quoted for 100 TSR-2s.

Had BAC been allowed to continue with the development and production of the

# TSR-2 PHOENIX OR FOLLY?

TSR-2, then the money spent on the programme would have been kept in this country. Alternatively, the overseas order would have drained Britain's resources even further. The government, of course, was interpreting the TSR-2 and F-111 figures to suit their own viewpoint. Whilst certainly those figures in respect of the F-111 are probably true, this is not so in the case of TSR-2. If, for example, one explores the full facts, one realizes that although we see a huge price tag for one TSR-2, it certainly does not represent the figure for which the Treasury would have had to write the cheque. Let us consider for one moment the amount of people all over the country that were involved, directly or indirectly in the TSR-2 programme. Consider also the amount these people were paid in salaries that in many cases included large amounts of overtime payments. Now consider how much tax these same people would have paid from their wage-packets, money that would have gone to the Treasury. In addition this can be taken further by considering the transport costs of each employee as he travelled to and from work. He would have to buy petrol for his car, on which he paid tax, and buy goods that would also eventually return money to the Treasury. This illustrates just one small insight into the 'invisible earnings' the Treasury could get back from a project for which they would have been writing out the cheque.

The next important consideration must be the government's action in cancelling TSR-2 and ordering the F-111. At the time an estimated 12,000 jobs were lost immediately in the industry. Some of the people affected chose to emigrate while others found new employment, but of course the majority fell into the ranks of the unemployed. The Treasury naturally lost all income from workers who emigrated, and although the ones who sought new jobs and found them would still have been contributing to the Treasury, those now out of work would of course be claiming benefit and that comes out of the Treasury purse. We now know that the government eventually cancelled the F-111, spending another £46 million on cancellation costs, a sum that in no way benefitted the taxpayer. As can be seen from these figures the claim that it was the exorbitant price of the TSR-2 that was the reason for its demise is now surely questionable and the real cost of cancelling the TSR-2 project will perhaps never be fully realised. As I mentioned earlier, this is an extremely intricate and difficult subject and my calculations are based on figures generally available from various publications. However, even if the figures are slightly exaggerated, there appears to be no conclusive evidence that could have led anyone to believe that the purchase of the American aircraft would have given the savings so consistently postulated by the government at the time.

**Electronics Industry Survives.**
However, it was not all doom and gloom emanating from the demise of the TSR-2. The electronics industry played a tremendous part in the programme and towards the end appeared to be in a stronger position than even the airframe manufacturers. Had the aviation and electronic industries perhaps become victims of their own advanced technology, or merely been the innocent victims of the demands of modern warfare? At one time, big may have been beautiful, but today's technology has progressed to the other extreme. As the demand for more efficient electronics has grown so has the need for equipment to become smaller. The weight thus saved also allows aircraft to carry either more equipment or fuel. Although the size of components has become smaller, the costs have increased considerably.

The role the industry was to play in the TSR-2 story was extremely important. Had it not been for such sophisticated equipment then the capabilities of TSR-2

# PEOPLE, PLANES AND PLANS

*A Tornado F-3 of 11 Squadron lifts off from Boscombe Down. Not necessarily an ideal fighter, the F3 Tornado demonstrates what can be achieved by adapting a common airframe.*
*(Rolls-Royce plc)*

would have been drastically reduced resulting in the aircraft being 'just another bomber'. In the event the electronics industry had in fact a considerable investment in this military project. However had they continued to depend on the British government for a constant supply of contracts for military equipment, then they would have faced bankruptcy very quickly! In fact Elliott Automation had anticipated the state the British aircraft industry was in and immediately after embarked on a huge sales campaign in the United States of America. Here, both the aircraft manufacturers and the military quickly recognised the huge potential of the equipment Elliotts were marketing, both in digital computers and head-up-displays. The USAF and Navy quickly realised the advantages of such equipment and duly re-equipped their aircraft, by replacing their electro mechanical systems with the latest revolutionary equipment from Elliotts. Meanwhile in Britain the MoD were extremely hesitant in taking advantage of the very components that they themselves had laid down as a standard in the TSR-2.

Ferranti, with their terrain-following radar, had worked extrememly hard in their bid to convince the government that the British radar system would be the ideal unit for the TSR-2. An extensive flight programme, using a Canberra bomber, had gathered data on the reflective characteristics of a wide variety of terrains and obstacles, particularly masts and towers. As the programme progressed and confidence grew, the flight testing aircraft was changed to a Buccaneer. This particular aircraft was the second development model and having completed its own flight test programme was due to be dismantled and scrapped. As it happened this aircraft turned out to be an excellent choice as it was the ideal airframe for the work demanded. On the other hand, though it was smooth, stable and extremely strong, it was not particularly popular with the ground crew, as it was an early type of airframe, powered by out of date engines and used obsolete parts which made spares hard to come by.

# TSR-2 PHOENIX OR FOLLY?

By the time cancellation came, Ferranti had gained a tremendous amount of experience, not only in terrain following, but also in other modes of the radar's operation. Fortunately the MoD allowed the company to complete most of its development programme, which included an exercise to investigate the effects of extremes, of hot and cold temperatures and vibration on equipment for long periods.

Other work included terrain avoidance, but perhaps the most lucrative spin-off was the possibility of technology used in the TSR-2 being used in the IDS version of the Tornado. This, unfortunately was not to be. Due to extreme political pressure, exerted by the German Government, the decision was made to use the American Texas Instruments terrain following radar. Therefore Ferranti again lost out, and had no opportunity to use their equipment. Ferranti continued their campaigns in an effort to establish an obstacle avoidance radar system for civil airliners. Again this came to nothing as the airline manufacturers felt that whilst such a system would be advantageous, it would not increase passenger revenue.

One of the team involved with the Ferranti system remembers quite well attending a lecture some years after the TSR-2 project was cancelled. This lecture was given by an American who was himself involved in the U.S programme of development by Texas Instruments for the F-111 radar system. He went on to describe the various problems the American company had encountered which Ferranti had not only anticipated, but had also successfully overcome much in advance of the Americans. This, I feel, illustrates perfectly the tragic waste of a tremendous amount of good, sound, British technology.

## A Phoenix in the Making.

In 1981, a member of the Conservative Aviation Committee commissioned a feasibility study to look at the possibility of rebuilding TSR-2 and an article appeared in *'Air Pictorial'* on this subject, written by Graham Wilmer. The article makes very interesting reading and I have highlighted a number of points. The study was based on the original design and airframes that were still in existence.

When the Conservative government initiated a review of the TSR-2, this is what the aircraft may have looked like had the recommendation gone ahead. *(Phil Kingham)*

# PEOPLE, PLANES AND PLANS

The report does however, recommend a number of changes, in particular, to the engine intakes. TSR-2 had semi-circular intakes with a half cone in order to control the intake of air at supersonic speeds. It was felt that simply altering the design of the intake to that of the intakes on the Concorde or Tornado, would both increase efficiency and at the same time give better airflow control. The article goes on to suggest that the Olympus 593, the Concorde engine, could be fitted to a rebuilt TSR-2. However, as desirable as this may be in terms of the advanced technology this engine provides, the larger diameter of the 593 by some 5 inches, would mean that a major redesign of the rear half of TSR-2 would be necessary. It is possible though that the technology gained in the years since cancellation could have been used to enhance the original 320 series engine used in TSR-2, thus making it one of the most powerful and efficient engines currently in military use.

The use of modern avionics would give the aircraft increased efficiency over the various systems originally used in the project by being able to handle the data with more control. At the same time this would give the airframe far more room to carry an increased volume of fuel or stores, which would give a greater range and on station capability. Another weight-saving device would be the use of materials such as Carbon Fibre. This material is strong but weighs very little compared to metals with similar strength. TSR-2 has many panels including the bomb doors and large undercarriage doors where the use of such material would be a tremendous

*The Eurofighter 2000. An essential weapon if Britain is to maintain an effective defence and policing policy. This aircraft, shown here at Warton, is the first British prototype prior to going to the paint shop. (BAe)*

weight-saving advantage. On many military and civil aircraft such material already forms a large part of the moving surfaces and TSR-2 with its large fin and tailerons, would no doubt have gained much from Carbon Fibre. The implications are immense as the new technology would have enhanced the TSR-2 considerably. The Secretary of State for Defence, Francis Pym, reviewed the proposal. It would have been a daunting task, and the economics of reviving the two airframes was far too expensive to consider and the proposal was turned down.

**Has Anything been Learned?**
What has the government and industry learnt from this debacle? Although the after effects of the Supermarine Swift affair played a significant role in the procurement of military aircraft, thereafter there is still evidence to show where the system continues to break down. After the fiasco of the TSR-2 the way forward appeared to be in collaborative projects, such as the Jaguar and Tornado, with our European partners. The Concorde had played its part in laying the foundations for such ventures. Now no single government can cancel the whole project but merely its particular role. Heavy penalties are imposed on those who choose to withdraw from such programmes.

There is continuing evidence however that these projects are still vulnerable in repeating mistakes made in the past. The Panavia Tornado project appears to have fallen victim to these similar mistakes. A great deal of time and money could have been saved had results gathered during the development of the initial phase in Germany, been carried over to the member partners when they reached that same stage. Instead each went about the same task its own separate way trying to rectify problems that had already been solved. Next is the Euro-Fighter. But a question mark still hangs over the viability of this project, particularly in the wake of Glasnost and the unification of Germany. Why?

Well, the break-up of the USSR has resulted in even greater instability throughout Eastern Europe and Asia. The weapons are still there but are now spread throughout areas of dubious political stability. We are already seeing the 'separate' nations charging up their nationalism. The nuclear proliferation will not go away, and as such the United Kingdom will continue to need a high quality defence and policing capability. This can only be achieved by using state of the art equipment. The mistakes made with TSR-2 must never be repeated, and therefore projects like EFA must remain. Defence cannot be switched off and on merely to provide a government with a stable fiscal policy. Because of the time required to develop effective and reliable equipment it must be a continuous commitment. We must never be caught out by believing that other nations will support either our principles or our defence.

It would have been hoped that during the time of radical change since the Second World War the government would have made the necessary changes within the bureaucracy of Whitehall and that such long term commitments are extended to future governments, instead of each party wasting time and money fighting over such important issues.

The various European collaborations virtually prohibit the government from intervening as it did with the TSR-2, but this is still not so with projects instigated from within Whitehall. Although the Ministry of Technology, so important a decision maker in such projects, is now part of the Ministry of Trade and Industry, we can still see where intervention has often led to blunders within the procurement process.

# PEOPLE, PLANES AND PLANS

*Nimrod started life as Britain's first jet passenger aircraft. Modified to take on the role of maritime patrol aircraft and later as Britain's attempt at an Airborne Early Warning aircraft. Was it scrapped because it could not do the work or was scrapped merely to satisfy a political struggle? (Rolls-Royce plc)*

## AEW's and Governments.

This type of intervention was particularly apparent with the radar system proposed for the Nimrod. In the 1970's Britain's airborne radar system was reliant on the outdated Shackleton and its equipment. It was evident that it was getting to the point where reliability was being impaired mainly because of the lack of spares, particularly for the aircraft. It was even rumoured that the RAF were robbing museum pieces in order to keep the Shackletons flying! The stage had obviously been reached where the RAF had to have, not only newer aircraft, but also a more up to date radar system in order to cope with modern day requirements. The system proposed by GEC appeared to be the one most likely to meet the RAF's needs, and, at the same time, give it a huge advantage over its nearest rival, the Boeing AWAC.

After considerable sums of money had been spent on developing the new radar, it was cancelled. Why? References made by an unfriendly media at the time, quoted stories of inefficient equipment, that failed to meet the required standards, and gave particular emphasis to the supposedly exorbitant costs of such a project. So much so that eventually the system acquired the reputation of no longer being a viable proposition.

Reading these references could give the impression that the company manufacturing the radar appeared to have inflated its costs so highly in order to produce huge profits to satisfy its investors. But is this the real reason, or merely media sensationalism, fuelled by misleading information from an inefficient government bureaucracy, in order to make the manufacturer the scapegoat. The true story of the radar, like that of the TSR-2, may never be told. However information regarding the whole radar project appears to be somewhat different when viewed from other angles.

The RAF did indeed require a new AEW system to replace the equipment in use in its ageing Shackletons. A tender was put out and various companies submitted their proposals. These proposals were submitted based purely on the information

## TSR-2 PHOENIX OR FOLLY?

set out in the operational requirement. The manufacturer of the radar had no say as to which aircraft his equipment was destined for, but merely received the parameters within which his equipment must operate. The requirement was for a radar that would be used for picking up contacts travelling in excess of 70 knots over the sea. The equipment was to make use of an airframe that was currently available. The airframe chosen was of course, the Nimrod. At first it appears to have been an ideal airframe in which to carry the radar, for at the time it was doing an excellent job as a maritime patrol aircraft. The origins of the aircraft design go back to the Comet which was introduced in the early days of jet travel, and around which the Nimrod was based. Although the Nimrod airframe had been considerably updated from that of the Comet, its major drawback was the lack of space when the radar system was installed. The aircraft's outline was changed considerably by the addition of huge bulges on the nose and tail. It was, however, within the aircraft that the real problems arose. Radar units of this calibre are not particularly small, and so when it was required to change a specific unit or module, the main passageway would become blocked by the operation. Not exactly an endearing characteristic in an aircraft that would have a high workload. In comparison, the Boeing AWAC, based on the Boeing 707 airframe, is somewhat larger and can even accommodate a coffee machine!

Nimrod's radar system, perhaps, like all new pieces of equipment did not at first perform to the required standard. However if one studies a map of the North Sea and its coastlines, you cannot help noticing the huge land masses which lie in all directions. This of course, is bound to have an affect on a system designed to function mainly over flat areas. Subsequently the specification needed to be

Is this the only way Britain can gain recognition for its technology? A further example of British technology having to be given up. The ultimate Harrier, the AV8-B seen here at Rolls-Royce Bristol undergoing engine trials. The aircraft is now built by McDonell Douglas in America. *(Rolls-Royce plc)*

# PEOPLE, PLANES AND PLANS

updated in order that the radar could allow for the change of terrain. For reasons unknown such modifications were forbidden. This meant that the system had to cope with an environment it had not originally been designed to cover. The radar returns would be somewhat different to those originally anticipated and give some confusion to the operator. By the time the project was ready for an official assessment it had already attracted a number of enemies who either felt the costs were rising out of all proportion, or who now considered that the airframe had become totally inadequate for the system.

When George Younger, then the Defence Minister was given a demonstration of the equipment by the RAF, it was with an early type system and even then the RAF had failed to implement certain changes as recommended by the manufacturer to overcome some basic problems. There were even unconfirmed reports that during major trials such as this, American AWACS in the vicinity flew with their radars washing over the Nimrod. This would drastically degrade the efficiency of the Nimrod's own radar and of course affect its overall performance rating in the eyes of the government. It is interesting to speculate on the reasons why the Americans should do this, if indeed it was true! Whilst on the subject of the American radar, was it any better? Again according to informed sources, possibly not. While information on equipment currently in use is restricted, it is extremely difficult to assess the two systems fairly. However, there was one significant difference. Apparently the American AWAC system is limited to tracking objects moving in excess of 150mph, and although not immediately obvious, this could exclude vehicles such as helicopters and other aircraft able to move slowly, in particular those of the Harrier type.

The purchase of the American AWAC went ahead mainly because there appeared to be no other alternative capable of conducting the required operation.

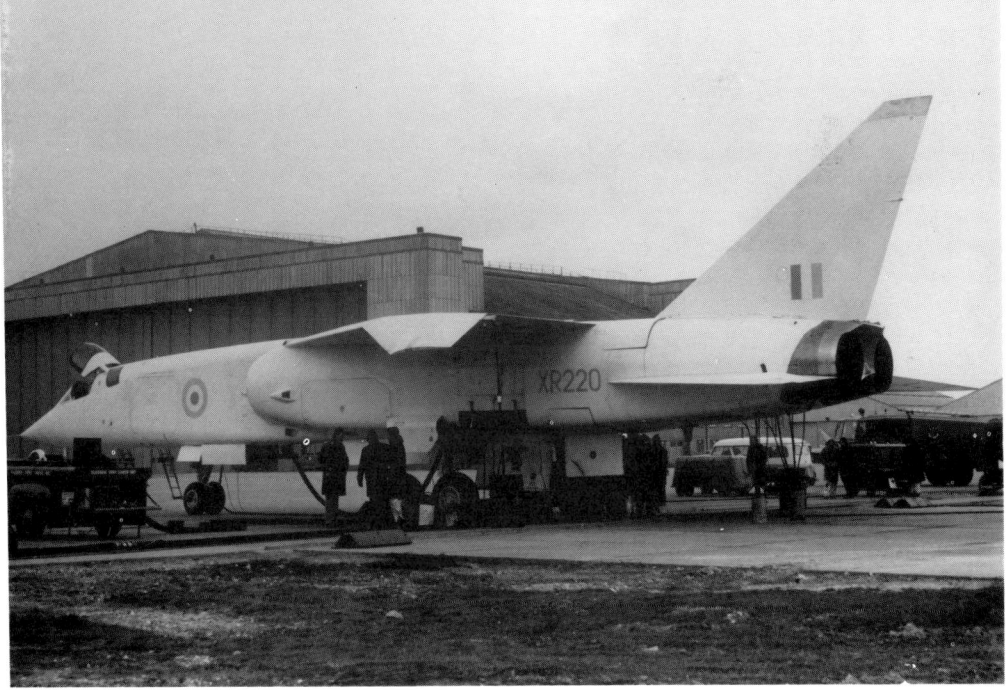

XR220 undergoing fuel-flow trial in late March 1965. *(BAe)*

The idea of producing a similar aircraft in conjunction with our European partners would, I feel, have too much against it. There appeared to be no requirement by other nation's within the community for such a system and the time delay would have been far too prohibitive.

**TSR-2 and Beyond.**
The whole question of defence is an ever changing one and is not helped by the differing views of the various political parties. To anticipate effectively the needs of the nations defences, in say, 10 years time, is exceptionally difficult, especially against a background of the constantly changing political fortunes throughout the world. But surely the wasting of money, technology and manpower cannot be allowed to continue, and in future closer cooperation between the political parties over this question must have a significant part to play. Many of the specialised companies involved in the TSR-2 programme made a number of technological breakthroughs in producing components that could also have had an immediate effect on current civil applications. In a number of instances, these were not allowed to continue because of the veil of secrecy surrounding the project. As a result, such products were either delayed, or, even worse, discontinued with any possible benefit lost forever. To allow such products onto the civil market at an early stage would have ensured continuity not only in its development and to some extent environmental stability, but would also have allowed some return on the huge investment of public money.

TSR-2 was the benchmark for future military aviation products and it is interesting to speculate on what would have been the state of Britain's military aircraft today had we gone ahead with the TSR-2.

Throughout this and other chapters reference has been made to the shortcomings of the F-111 when compared to the TSR-2. Obviously the comparison has had to be made because of the Government's intention to order the F-111 as a replacement for the TSR-2, but in no way should it be concluded from this that the F-111 in my estimation was, or is, an aircraft of dubious quality or effectivness. Whilst it is true to say the F-111 started out dogged by many problems, the might of General Dynamics went on to ensure its final product was a superb aircraft. I personally feel that the two aircraft were designed to perform different roles and therefore one of them is always going to come out badly if there is an attempt to fit it to the others role.

# CHAPTER NINE
## The Penultimate Chapter.

During the period between February and April 1965, the TSR-2 programme had gone from strength to strength, with time lost during many of the delays caused by mechanical problems, having been clawed back. The aircraft was due to be laid up during April in order to carry out many planned modifications. The problem caused by the fin jack failure meant that the scheduled lay-up was brought forward and at the same time a number of minor faults that had been exposed during the test flights would also be corrected. The flight test development programme had made some huge advances helped by the fact that the aircraft's stability and control had proved to be of an ideal standard, with the airframe showing tremendous stability, particularly in those areas where problems had originally been anticipated. So, all those involved were extremely confident that the programme could now achieve all of the targets laid down in the revised schedule.

There is no precise record of the destruction of the TSR-2. When the cancellation came the build programme was particularly well advanced, there being seven airframes in the final stages of manufacture, from XR221 through to XR227. The first two aircraft, XR219 and XR220, were, of course, already flying or about to fly. XR221 was almost ready, having flight engines fitted in order to carry out the integration trials. The contract for the pre-production models, XS660 to XS670, had been signed in June 1963 with the building programme having already been authorised and once again at the time of cancellation, this programme was particularly well advanced in respect of the production jigs.

Many articles and books written at the time either comment on the somewhat indecent haste or even the scandalous speed at which the destruction of the project was carried out. BAC claimed that this was necessary to prevent people from indulging in any unauthorised working on anything that was left of the project,

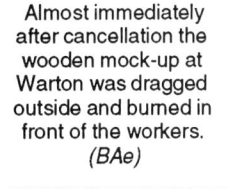

*Almost immediately after cancellation the wooden mock-up at Warton was dragged outside and burned in front of the workers. (BAe)*

# TSR-2 PHOENIX OR FOLLY?

saying that *"it would be quicker and easier to restore morale in the workforce if there were no reminders!"* All books that referred to the technical aspects of the TSR-2 were ordered to be destroyed by the MoD, and many of the photographs that were taken of the aircraft, both during its run-up to first flight and whilst flying have been systematically destroyed for reasons that still remain unknown. The test establishment at Boscombe Down, claims to have no official records of the flights made from there. It was intimated that such documents appeared to have 'been deliberately lost'. This I found most unusual as the establishment keeps complete records of all such flights made by experimental aircraft from Boscombe, whether on service trials or industry testing.

**Scrap it!**
Very soon after the cancellation was announced, BAC was ordered to reduce to scrap, all the airframes and associated equipment. The wooden mock-up of the aircraft built at Warton, was dragged outside and burnt. The Ministry of Technology had invested over £12 million in specialised tools and jigs specifically for the production of the aircraft, and these were also immediately scrapped even though some sources say their scrap value was only around £50,000. Most of those airframes on the assembly lines at Weybridge and Salmesbury were immediately disposed of, and there are numerous pieces of photographic evidence to support this, although this itself appears to cause some confusion. One particular photograph which appeared in 'Flight' magazine, shows the rear half of XR226 with a caption stating that it was taken at a scrapyard in West Bromwich. Another witness sent me a photograph that he himself had taken in the famous R.J.Coley scrapyard at Feltham, Middlesex, sometime during 1965. It depicts a large number of Hawker Hunter fuselages, but just to one side one can clearly see the distinctive shape of TSR-2 fuselages. The witness claims that these had the serial numbers of XR226 and XR227. It is most unfortunate that the photograph itself does not confirm such a claim. It is true however, that the Coley yard did receive many of the front sections built at Weybridge, together with a number of rear halves that had been sent to Weybridge for final construction. In September 1965 pieces of fuselage from XR224, XR225 together with XR226 were again supposedly seen

Taken at the Coley scrapyard in Middlesex sometime in 1965, the picture shows a large number of Hawker Hunter fuselages, but also to the right the distinctive shape of the TSR-2 fuselage can be seen.
*(Peter Nicholson)*

# THE PENULTIMATE CHAPTER.

here. Most of the other rear halves built at the Lancashire factories, were first dismantled at the Salmesbury works before being moved to a scrapyard in Yorkshire. It was also rumoured that certain components were seen in a scrapyard near Chesterfield, Derbyshire, shortly after the cancellation period, but there has been nothing to substantiate this and therefore it cannot be confirmed.

**MoD Shoeburyness.**
A number of airframes were retained for use by the MoD at various test establishments. The Proof and Experimental Establishment at Shoeburyness is one of the bigger sites run by the MoD, and has ranges here extending out to Foulness Island. It may be remembered that this was to have been the area where the proposed second London Airport was to be sited on the nearby reclaimed Maplin sands. However, following a series of protest demonstrations by various groups, the whole project was dropped and the Army, Navy and Air Force continued to do their testing there. Between April 1968 and July 1969, a number of publications made reference to the airport project and featured articles on the test establishment, which showed various photographs of TSR-2 XR219 still complete and awaiting trials. On the northern part of the island, the Royal Air Force has an area available for destructive testing using old and usually 'out of hours' aircraft that can serve no other purpose. The central area is known as White City, as the buildings in the vicinity are painted white. It consists of technical and administration buildings plus a large open hard standing area where the aircraft are parked and prepared prior to being moved out to one of the trials areas.

The first TSR-2 airframe sent to Foulness Island was XR221, on the 13th September 1965. The movement card held at RAE Farnborough records this aircraft as having had engines fitted at Weybridge in November 1964, and then further replacement engines in April 1965. It was therefore essential that all equipment, including the engines, be removed prior to the transfer. The transportation of all the airframes being moved to Foulness Island was done by a Messrs Annis & Co, heavy haulage contractors. The first component to arrive at the main-gate in Shoeburyness was the fuselage, and it was then to be transported out over the Havengore Bridge to Havengore Island then on to Foulness and finally to White City, a distance of some seven miles from the establishments main gates. The fuselage created few problems, but it was to be a different story when the main wing arrived on the 20th May.

As with all the TSR-2 wings shipped to Foulness, they were bolted to custom-made angular steel frames but, because this made their height and width greater than the original Havengore Bridge, a plan was devised to transport the wings over a different route. It was therefore decided to move them to White City over the 'Broomway'. This was a natural roadway across the hard sand lined with wild broom and led out across the lower sands, which were accessible at low tide. The transporters would be escorted by military personnel in DUKW's. The vehicles would come in from the seaward side and over the sea wall via a roadway at a point close to White City. This apparently straightforward task was made very difficult when the long low-loaders tried to negotiate the sea-wall. The low-loaders were getting stuck negotiating the apex of the sea wall and had to be chocked up with the tractor removed. Following this the trailer was lifted off the chocks and inched forward by using a crane lorry with slings attached to the front of the low-loader. This manoeuvre was repeated until the low-loader was over the sea wall and down the slope where the tractor could then be recoupled and the rest of the journey

# TSR-2 PHOENIX OR FOLLY?

completed. Even then there were certain trees that had to be drastically trimmed and various road signs that had to be dug up to allow the low-loaders to negotiate the narrow winding road to White City. The staff were faced with this same problem for each wing sent to Foulness. Although the undercarriage was fitted in XR221 it was stowed in the up position for transportation purposes and so the fuselage had to be trestled to allow the undercarriage to be lowered later.

The next airframe at White City was XR223, along with the taileron, fin and wingtips on 27th September 1965. This was an uncompleted airframe, which did not have the undercarriage fitted. The main undercarriage and nose legs arrived later on the 5th October 1965.

*The difficulties of moving the TSR-2 airframes around Foulness Island were overcome by using a method of travel, that today, would be frowned upon by the safety administration. It was, claims those involved, one of the most effective and safe methods of moving such large components around the narrow twisting roads. (Phil Kingham)*

## XR219 Arrives at White City.

XR219 was the next aircraft sent to the establishment, but before describing its arrival and subsequent destruction, there are a number of facts worth detailing. As mentioned earlier, XR219 had last flown at Warton on March 31st 1965 with Jimmy Dell at the controls. Then after Don Knight's abortive attempt with flight number 25 in early April, the aircraft was laid up for repairs and important undercarriage modifications to be carried out by BAC. It had been hoped to retain XR219 at Warton after the cancellation in order that it could continue flying duties. This would have allowed the immense amount of data so far collected in the project to be finalised and the findings used in future developments. (In fact this data was used in the Concorde project). In Chapter 7, mention is made of the government's views on the continuation of flying TSR-2, and although they had not actually refused, they had effectively vetoed the idea by insisting that the development costs of all future flights must come out of the cancellation bill. There seems little doubt that XR219 was still capable of flying as the picture taken during a Warton Open Day on 15th July 1965 shows. This date can be confirmed from the records at Warton on Lightning production, which state that the F3A Lightning, also in the

# THE PENULTIMATE CHAPTER.

Prior to XR219 being taken to Shoeburyness the airframe was displayed during an open day for the Royal Aeronautical Society at Warton. The date is sometime after June 1965 which is confirmed by the F3-A Lightning which had been delivered to Warton on June 25th 1965. *(BAe)*

picture, had been delivered from the Lightning production line at Samlesbury the previous day. When BAC were told of the intention to send XR219 to the Proof and Experimental Establishment at Shoeburyness, they had put forward a proposal to fly the aircraft to Southend Airport, which was the nearest airport to the establishment. However, it appears the government quickly dismissed this idea, letting it be known they had no intention of seeing any TSR-2 in the skies again.

Toward the end of July and in early August 1966, the airframe was prepared for its journey to Foulness Island. All equipment was removed including ejector seats, engines, and radio. Most of this equipment was to be sent to Cardington or Henlow. In early August a low-loader carrying XR219s wing started for Shoebuyness. On August 10th the vehicle arrived and the difficult journey of getting the wing over the headland began on August 12th. The following week on 17th August the fuselage, associated panels, fin, taileron and wing tips arrived at White City. As with the previous fuselage belonging to XR221, the undercarriage had been retracted and locked so the fuselage again had to be trestled. At the time no trestles had been supplied to Shoeburyness from BAC so P&EE had to manufacture some specially to support the fuselage before the undercarriage was lowered. To lower the undercarriage an electrical circuit had to be made and used in conjunction with a Silcodyne hydraulic test rig. So far all TSR-2 equipment had been sheeted for the journey to Shoeburyness, but when XR219 was unloaded and trestled it was left uncovered and at the mercy of the elements.

During October 1966, a large amount of extra aircraft and ground equipment was sent to Foulness from Weybridge and Warton. This included three 25 ton Ski-Hi jacks, a wheeled towing arm with an additional tubular section, a pair of tall wing trestles, a Silcodyne Hydraulic service rig and a hydraulically operated cantilever lifting platform. The Silcodyne test rig would be needed to lower the undercarriage on XR219 but there was only a limited supply of the Silcodyne fluid. During 1989 the 25 ton jacks were overhauled at Shoeburyness and put back into service,

# TSR-2 PHOENIX OR FOLLY?

*The second of the Bristol 188s at Shoeburyness after cancellation. In the background is XR219 being manoeuvred by a Land Rover. (MoD)*

hopefully to give another 25 years service. The towing arm has been modified slightly to fit Buccaneers at the establishment, and this allows the aircraft to be moved round with relative ease. The cantilever platform still performs well, being particularly useful in moving aircraft and large pieces of equipment around in tight areas.

### XR219 is Readied for Tests.

In September 1966, RAE Farnborough requested that XR219 be rebuilt to accomodate certain forthcoming trials, and so work began on a rebuild. At the time, the personnel at Shoeburyness were involved in the rebuilding of some of the Valiants sent there shortly after the type had been prematurely withdrawn from service and were doing trials on these particular aircraft. Initially owing to the onset of bad weather, the lowering of the undercarriage and other essential trestling was the only work done, but by mid 1967, the rebuild of the aircraft had virtually been completed. Final completion in fact was achieved on the 6th of May 1968, when the flaps and tailerons were the last pieces to be fitted. Although instructions had been given back in 1966 for XR219 to be rebuilt, the aircraft was to stand out in the open for some time and during this period a number of attempts were made to ensure that the only TSR-2 to have flown would be preserved. The staff at White City had even tried to save it themselves by ensuring that XR219 was put beyond the pale. However, this never happened and in April 1973 the Physics Department at RAE Farnborough issued a programme of trials work to be carried out on TSR-2. XR219 was prepared for the trials with the emphasis being on the integrity of the fuel system in the wings and fuselage. The northernmost range was to be where the trials

# THE PENULTIMATE CHAPTER.

were to take place, a distance of three-quarters of a mile from the White City area. Because of the narrow and twisting roads on Foulness the airframe could not be towed with the custom made towing arm supplied to P&EE. Therefore an intricate system was devised to carry the aircraft to the range. The job of fitting the main wing had proved quite difficult and it was therefore decided to remove only the tailerons and fin.

Two Smiths crane lorries were positioned, one to the front and the other to the rear. The airframe with the undercarriage still in the down position would be lifted some 3 feet off the ground. The rear 30 ton crane lorry would then travel in reverse with slings fastened round the taileron stubs. The front crane travelled forwards with a sling round the fuselage underneath the cockpit area. The whole procedure was overlooked by a controller to ensure that both crane lorries moved at the same walking pace to ensure they were kept upright and that the front sling did not slip off. Having visited the establishment in 1988 I can appreciate the problems involved with the narrow roads and twisting corners. During my visit a small crane lorry had ventured on to the soft verge during an overtaking manoeuvre and as a result the vehicle had slid into the ditch. According to the personnel there, the lorry would simply have to stay there, as any attempt to recover the vehicle would probably result in the recovery vehicle toppling in as well! This method of moving heavy equipment around Foulness would certainly contravene the Health and Safety regulations of today. However, I was assured by the controller at Foulness, that although this method may appear somewhat dangerous, it was a far more reliable, efficient and cheap method of moving equipment around than is sometimes the case today.

A number of photographs exist of TSR-2 XR219 at Shoeburyness. The condition of the airframe deteriating over the months.
*(MoD)*

# TSR-2 PHOENIX OR FOLLY?

**Mishap on the Range.**
After the journey out to the North Range, the tailerons and fin were fitted and the aircraft was prepared for the forthcoming trials. However, it seemed that XR219 would not die without a fight. During the night before the day of the trials, and without warning, the starboard tandem bogie sheared from the main leg. On the following morning the trials staff moved out to the range ready to commence the trials, only to find XR219 listing heavily to starboard with the main undercarriage leg deeply embedded in the concrete. The trials had to be delayed whilst the aircraft was trestled in the upright position. There was some concern as to why the undercarriage should have sheared without warning, so a memo was sent to all TSR-2 holding units warning them of the problem. It is believed that the severe oscillation in the undercarriage during the landing phase had been the major factor in the failure. However, as XR219 was the only TSR-2 to fly, it was felt that no other similar aircraft would have this problem.

Although Shoeburyness claims a photograph exists of XR219 on the range with the undercarriage collapsed, it has so far not been found. Phil Kingham, again working from eye witness accounts has depicted XR219 out on the range with the gear embedded in the concrete. *(Phil Kingham)*

**Trials Begin.**
Information as to what the trials were all about is, even to this day, still classified, but from snippets of information gathered during my visit to Shoeburyness in 1988, it appears the trials work was purely to establish how the airframes would stand up to constant gunfire.

The RAE at Farnborough were prepared to say that in the main, the fuselage sections were primarily used in trials to establish the extent of external lethal blast contours and this was done by using pure blast warheads to attack the structures. Some were used purely as targets in lethality trials to test the effectiveness of new and experimental warheads intended for anti-aircraft and ground-to-air missiles, whilst others were being used simply as targets in gun fired projectile lethality trials. The wing sections were used for similar testing designed mainly to observe the damage sustained when attacked by various fragmenting warheads. TSR-2's structure at the time was considered to be quite unique as regards the materials used in the construction of the aircraft and this gave the RAE an ideal opportunity to assess both the construction method and materials used, so that comparisons could be made with aircraft structures and materials previously used.

# THE PENULTIMATE CHAPTER.

*XR219 prior to being taken out to the northern trials area. (MoD)*

The trials on XR219 continued right through until May 1975 and it appears the fuselage took the brunt of the pounding. After these trials, what was left of the aircraft was dismantled on-site. The fuselage was moved back to the White City area using the same procedure that had taken it out to the range. Barely recognisable as a fuselage and incapable of being of further use, it was disposed of as scrap in January 1977. The main wing was returned to the hangar at White City and repaired sufficiently to enable it to hold fluids ready for further testing, which was done regularly on this particular component right up until 1982, after which it could serve no further purpose and was duly scrapped.

### Further Trials Work.

As mentioned earlier, XR221 and XR223 were received incomplete from the manufacturer and therefore it was not possible to rebuild these airframes to the same standard as XR219. However various trials of a similar nature were carried out on the fuselage and other components from XR223 between November 1968 and August 1970, and on XR221 between February 1970 up until August of the same year. All of the testing was done on the north range and the fuselages were moved to the site using the same methods that had taken XR219 out. This time though, there had been no main wings fitted and the undercarriages had been locked in the up position.

## Olympus Engines.

During September 1988 I visited the site to discover exactly what TSR-2 pieces still existed on Foulness Island, but of the equipment mentioned earlier, very little remains. By far the biggest find were the Bristol Olympus 320 engines. After the cancellation of the project, a number of firms and establishments had engines in various stages of manufacture. During November 1966, P&EE received two Olympus engines and jet-pipes from the MinTech, Sevenhampton in Wiltshire. These were in a 'finger tight' condition indicating they had been the remainder of a production line contract and hurriedly bolted together before being moved to Shoeburyness. In January 1968 another unit was sent from Bristol Siddeley, number 22210. This was a pre-production unit that had been originally fitted to XR219. Another Olympus was sent from the National Gas Turbine Establishment at Pyestock, Wiltshire on 7th March 1968, and a further seven were sent from Dowty-Rotol on the 22nd May 1968. All these engines were housed on the engine dump and in various states of repair, one having evidence of experimental work. From sources at Shoeburyness I found it was not uncommon to mount an engine inside any fuselage type, start it up and then use it as a ballistics target for munitions vulnerability trials. However, no engines of any type were fitted to the TSR-2 fuselages. Some engines were on the correct type stands, one was still boxed hiding in the undergrowth, and another was so overgrown with brambles, that it was impossible to find an engine number. Alongside the engines were a pair of Olympus AB233 jet-pipes laying on the correct mounting trolleys and a short loose unit. In 1988 all of this equipment was still the property of RAE Farnborough, however, I understand that in February 1989, Brian Trubshaw moved a number of the the engines to RAF Little Rissington.

There follows a list of engines that were at Shoeburyness in 1988, but are now believed to be at Little Rissington.

*Apart from airframes, the establishment at Shoeburyness received a large number of the TSR-2 Olympus engines. Most of these engines have been saved and are believed to be at Little Rissington... perhaps some-one can confirm and advise the publishers. (Author)*

# THE PENULTIMATE CHAPTER.

>Bristol Siddeley Olympus 320X  22214.
>Bristol Siddeley Olympus 320X  22216.
>Bristol Siddeley Olympus 320X  22219.
>Bristol Siddeley Olympus 320X  22220.
>Bristol Siddeley Olympus 320X  22226.
>Bristol Siddeley Olympus 320X  22227.
>Bristol Siddeley Olympus 320X  22230.

**Shoeburyness Today.**

Out on the Midway dump were two main wings, and according to the records at the Establishment, these are part of a consignment received by P&EE between February and March 1980, when RAE Farnborough sent a main wing and BAC sent four. Only one wing carried markings and that was from XR224. In the canteen area of the White City technical buildings there is an amusing photograph of a test that apparently went wrong. It shows an aluminium ladder sporting a bullet hole. It had evidently been placed a little too close to a ballistics trial intended for the wing belonging to XR224, that was hanging from a crane in the background. Of the other four wings, only two still exist. It is apparent that one has been used for some ballistics trials as it has several holes of varying sizes, but the other, apart from moss and brambles growing over it, appears in good condition and will be retained for further trials work. A vast amount of aviation equipment has passed through Shoeburyness since TSR-2 and the scrapyard there is regularly visited by scrapmen removing equipment that is no longer needed. The chances of finding any more TSR-2 pieces now must be almost non-existent, although quite recently whilst clearing a ditch on the island, a propeller and nose spinner was found, believed to be from a Spitfire.

Two TSR-2 wings are still at Shoeburyness today. This one is at the Half Way yard. In the background are various airframes used by the establishment.
*(Author)*

# TSR-2 PHOENIX OR FOLLY?

*Sets of TSR-2 wings at the MoD test establishment at Pendine, South Wales in 1989. There are no records available to confirm their origin. (Author)*

## MoD Pendine.

At Dyfed, South Wales, close to the coast is Pendine, another Proof and Experimental Establishment belonging to the MoD. The sands here were famous in the 1920's for the record breaking runs made by Parry Thomas in his car 'Babs', but the sands now serve as a firing range for the military. During the late 80's when the Lightning was stood down from RAF service a number of the airframes were sent to the establishment to be used as balistics targets. In 1989, I visited Pendine to photograph these hulks and during my searches there a number of TSR-2 wings were discovered. There were no records available as there had been at Shoeburyness and therefore it was impossible to verify the origin of the wings or that the establishment had received any other TSR-2 pieces, especially any fuselages. The staff at Pendine had even been unaware of their presence and were unsure of the type of aircraft the wings were from! In all there appeared to be two and a half wings. All had been broken in half, though one still retained the leading edge, but both bore strong evidence of heavy experimental work. Again, neither wing had any identifying marks, (apart from the ones used by the establishment), but the possibilty that these were the wings used at Farnborough for fatigue tests is not improbable. A search of the site revealed no further TSR-2 equipment.

## RAE Farnborough.

Although RAE Farnborough had been deeply involved at the development stage of the TSR-2, very little equipment is actually retained there, the biggest exception being a forward fuselage. This was housed out on the northern open dump. The section itself appeared to have been used in experimental work as it bore traces of burning from both the cockpit areas. Farnborough were unable to supply information as to its origin. In August 1992 the section was moved to the Brooklands museum in Surrey. The simulator at Farnborough, mentioned earlier, in which the pilots and navigators spent many hours training, appears to have melted into oblivion. It is said the unit was retained by the establishment in order to run various comparisons between the TSR-2 and F-111 nav/attack systems, but this also was unconfirmed. There was mention of the unit being used to train crew for F-

# THE PENULTIMATE CHAPTER.

*Brand new TSR-2 airframes being dismantled by scrap merchants at Salmesbury. Most of these airframes went to a yard in Yorkshire. (BAe)*

111 development, but because the two cockpit layouts were completely different in TSR-2 and the F-111, it was obvious such a proposal would not have worked. It is therefore quite possible that equipment from the simulator would have been used elsewhere in other projects and that eventually the basic framework would have ended up merely as scrap. During a conversation with Farnborough, someone made mention of a model of TSR-2 which they had, that had been modified to house a new radar system. The nose section of the model had been altered and out of the side was a dome, which on a real aircraft, would have covered a radar scanner. The person who had told of this was highly sceptical of such a development, pointing out that the drag factor alone on such a device, would have been so prohibitive as to make it totally unviable.

## XR220 Retained.

XR220 was at Boscombe Down and being prepared for its very first flight on the actual day of cancellation. After April 6th, the aircraft was retained at Boscombe for engine noise trials and detuner testing. The trials were carried out by Bristol Siddeley Field Service Engineers. Concern had been raised about the likelihood of structural damage to buildings by the noise from the Olympus engine. (It was of course essential this matter be cleared up before Concorde started flying). The tests went on well into 1966 and eventually finished in the December. No evidence was found that the noise levels reached would cause structural damage, or present a health problem to people living within the vicinity of the airfield. It should also be noted that noise levels were never a problem during the test flights of XR219. After the trials the aircraft was moved by road to RAF Henlow. Here all equipment was removed from the airframe, including engines and radio. Prior to the instrumentation being removed the wires themselves were also cut which would prevent the aircraft from being made airworthy without a tremendous amount of work. Henlow retained the airframe until 1975 when it was decided to put the aircraft on display at an RAF museum. The museum chosen was at RAF Cosford and in September 1975 the Aerospace Museum received XR220 and eventually the aircraft was put on display in a hangar quite unique in that it is dedicated to experimental and prototype aircraft. A number of other pieces of equipment were also donated to the museum, including an ejector seat, nose leg and a cutaway section of the large landing wheel and brake assembly. Cosford also have a superb model of the aircraft donated by BAC. A number of Olympus 320 engines are also at the museum but bear no identification and one has been sectioned to allow the inside to be viewed.

During the late 80's when it was decided to reduce the holdings and consolidate Henlow and Cardington, a large amount of surplus TSR-2 equipment was sent to Cosford in a number of large boxes. Much of this equipment appears to be the specialised nuts, bolts and screws used in the manufacture of the airframes. Sufficient instrumentation was found to allow reconstruction of the front and rear cockpits, to almost their entire and original specification. Further investigations uncovered the assembly that controlled the moving map display along with an actual terrain following radar unit. Perhaps one of the most significant finds was a

*XR220 was retained at Boscombe Down to assist with development of the Olympus engine for Concorde before being moved to Henlow. (BAe)*

*The Aerospace Museum at RAF Cosford was selected as the museum to receive XR220. This shows XR220 shortly after its arrival there. (Cosford)*

# THE PENULTIMATE CHAPTER.

digital computer manufactured by Elliott Automation. It is hoped eventually to have this equipment on display in surroundings that would demonstrate the role such equipment was to play in the aircraft itself.

### XR222 at Cranfield, then Duxford.

XR222 was another incomplete airframe on the assembly line at Weybridge when the end came, and although duly condemmed as scrap, it was transfered to the College of Aeronautics at Cranfield, Bedfordshire in October 1965. At the college, the airframe was used in post-graduate teaching. XR222 was to stand outside for all of its time at Cranfield and, with many panels missing, the airframe gradually started deteriorating. However, it was at Cranfield that the public had its first opportunity to see the TSR-2 at close quarters when, during an air pageant in September 1973, XR222 was put on display. Eventually it was decided to release this airframe to a museum and in 1978 XR222 was transfered to the Imperial War Museum for display at their Duxford Airfield. In 1983, british Aerospace agreed to replace some 60 missing panels and fairings. The following year, using XR220 at Cosford as a pattern, a team of apprentices and Instructors from the BAe Preston Training Department manufactured and fitted most of the panels. With this additonal protection, XR222 was again moved outside to face the elements.

During 1988, following an initiative from the author, the Imperial War Museum's technical staff made a detailed survey of the airframe. This resulted, early in 1989, in the start of a year long, deep conservation programme. Although the overall condition of the aircraft was relatively good, the wing leading edges were severely corroded. They required many hours of hard work by an enthusiastic team of Duxford Aviation Society volunteers led and guided by two museum technicians. The final result was an aircraft looking better than original. In February 1990, Roland Beamont was invited to Duxford for the roll-out of the now restored aircraft, and, after an inspection of XR222 he said that he had never expected to see a TSR-2 in such marvellous condition. Later in the year XR222 was moved to Duxford's new Super Hangar.

### Parts of TSR-2 at Coventry.

A few miles outside the city of Coventry, lying in the shadows of the old Armstrong Whitworth works, is Coventry Airport. Adjacent to the airport is Coventry's flourishing Midland Air Museum. In a corner of the museum lies parts of an aircraft

*The Institute of Aviation at Cranfield were most fortunate in having a TSR-2 airframe allocated to them. This shows XR222 having been rebuilt, but was destined to spend its time at outside at Cranfield, before being moved to Duxford. (Cranfield Institute)*

# TSR-2 PHOENIX OR FOLLY?

*On the move. XR222 being moved to the Imperial War Museum at Duxford. (Cranfield Institute)*

looking as though they were recently retrieved from an air crash. They are in fact parts from the TSR-2 project. The largest piece is a rear cockpit canopy alongside a Martin-Baker ejector seat. Other pieces include a control column and foot pedals, and tucked away in the bowels of an Argosy is a front wheel. How the bits came here no-one really knows, but it is possible they came from another museum. Eventually they hope to secure a corner within the museum and illustrate where the pieces fitted within TSR-2.

## RAF Teaching uses TSR-2.

At RAF Halton, the No-1 School of Technical Training, are various pieces of TSR-2 and following an invitation, I visited the station to discover what they had. It had always been thought that the largest piece of TSR-2 equipment at the school was a front cockpit canopy, but in the school's engine museum was an Olympus engine, typical of the type used on TSR-2. After confirming the engine serial number, OL2201, further investigation revealed this to be one of the early 22R development engines that had been used for test bed work. (Engines used in development on test beds had four digit serial numbers, whilst flight engines had five digit numbers). Displayed alongside the engine was a jet-pipe complete with afterburner.

The front cockpit canopy proved to be in generally excellent original condition with the exception of some small work done on one side. This took the form of some small holes that had been drilled into the cockpit surround. The actual perspex of the canopy was in superb condition and confirmed an earlier report that this particular piece was regularly polished to ensure the surface remained relatively scratch-free. The lamination in the glass had so far not 'hazed' and showed no imminent signs of

# THE PENULTIMATE CHAPTER.

*At RAF Halton, the No1 School of Technical Training, many pieces of the super bomber are still used as teaching methods in training new recruits, including this TSR-2 cockpit canopy which is still in excellent condition. (Author)*

doing so despite the passing of 25 years. A number of other pieces held by the school included a Dunlop brake unit, whose condition showed it had not been used on an aircraft. Another interesting piece was a starboard Power Control Unit and this would have activated the starboard taileron. Its condition was immaculate and even though the school's students regularly handled the unit it had stood the test of time. Considerable interest was expressed over a set of foot controls which had been in regular use in demonstrating the hydraulic braking system within an aircraft. The instructor told me he was unsure of the pedal's origins, but on closer inspection and comparing them with photographs of the TSR-2 cockpit area, I was able to confirm that they were indeed, the foot pedal controls for a TSR-2.

It was also said that a TSR-2 fuel gauge was here, but after a thorough but fruitless search of the base, the staff felt quite disappointed when they were unable locate it. All this equipment is still used by the school for serious instructional purposes and because the technology was so far ahead of its time the equipment will be in use for many years to come.

**TSR-2 Today.**
As far as I have been able to establish, this is all that has survived of what promised to be a Super Bomber. However there is nothing to say that the aircraft breakers, who had been instructed to reduce the airframes to scrap, have not somewhere along the way, collected a momento of their deeds!

In November 1982 Jimmy Dell visited Duxford and subsequently presented the control column from XR220 to the Imperial War Museum. In August 1991, Jimmy very kindly donated the control column of XR219 to the Aerospace Museum at

# TSR-2 PHOENIX OR FOLLY?

Cosford. Shoeburyness were sure the fuselage from XR224 was moved to Henlow, but so far this is unconfirmed, especially as a photograph exists showing parts of this fuselage in the famous R.J.Coley scrapyard at Hayes in Middlesex.

When governments speak of costs and saving money, it seems they place little value on British technology. Without doubt TSR-2 became the most expensive and technically advanced aircraft in the world, but was ultimately destined to become just a ballistics target. One hopes the information gathered from firing shells at the various components was of sufficient benefit in establishing new horizons in the future of aircraft manufacture.

After a year in the conservation hangar at Duxford in 1989, XR222 emerged looking immaculate.
*Below:* Pieces are still being uncovered. This cockpit section shown here at RAE Farnborough, has now been moved to the Brooklands Museum.
*(RAE Farnborough)*

# CHAPTER TEN
## Conclusion.

This, then, is the confusing story of one of the most controversial aviation products Britain has ever produced. When I first began this book little did I realise the sheer scale of controversy that lay behind an aircraft of such aesthetically pleasing lines as the TSR-2. Even now it is difficult to reach any firm conclusion because at the time there were so many changes happening in both the government and the industry. It appears these factors combined literally to bog the project down to the extent of causing the maximum confusion before the final agony of cancellation. Perhaps it would make things easier if I categorise those points which had the most influence on the programme.

- First and foremost, the TSR-2 was to be the most technically advanced military aircraft in the Western world. As such there was no accurate method by which the project could be assessed to determine its comparable costs. This very often led to misleading conclusions being made purely because the programme was being evaluated alongside projects that had nothing in common with the TSR-2 itself, either technologically, or in the methods by which the aircraft was being produced.

- Those against the project were by far the most prominent. Those elements of the media whose privileged positions allowed them to decry the project without question and what is more, without any official detailed knowledge of the

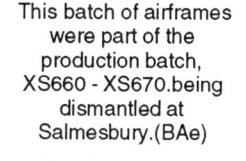

This batch of airframes were part of the production batch, XS660 - XS670.being dismantled at Salmesbury.(BAe)

subject, were perhaps the most influential when it came to forming the public's opinion of the TSR-2. There is no such thing as an unbiased newspaper and even they appeared to have a free hand at manipulating the mind of the public without reasonable consideration of the damage and effect it would have on the thousands of workers and their families who were involved in the project. Neither did they seem to consider the industry as a whole and the effect they would have on the future policy of the RAF or, the companies involved's prospects for future export business. Although the articles they published were not responsible directly for the cancellation, they certainly helped to turn public opinion against the project and the industry so that when the government took steps which showed them to be in agreement with the media, it gained them favour. Naturally any industry will have its critics, but such criticism should at least have a constructive element and be based on fact and not hearsay.

- Then there were those in a higher authority whose rank allowed them to display a bias that can only be described as unprecedented. It was these men who openly castigated, not only the project, but also the industries involved, accusing them of living off the fat of the land. Mountbatten, in his role as Chief of Staff, claimed he was totally unbiased in his attitude to the TSR-2 project and that his outspokenness against it was based simply on the cost of the project. However, the ferocity with which he attacked and condemned the RAF's project, together with his attitude over the Australian sale, demonstrated just how biased he was in wanting to protect his navy. Mountbatten was concerned not only that the project would place the nuclear capability firmly in the hands of the Royal Air Force, but would also drain the MoD of resources and that the navy would be unable to acquire the new through deck carriers. Together with Zuckerman, as his ally, the duo made a formidable force within the corridors of Whitehall. The American influence on Zuckerman appeared to be such that he catagorically refused to accept that Britain, as a nation, could produce an effective weapon that was also an exportable commodity within the western world. Then, of course there were always those in government...

- The fundamental policies of the Labour party appeared to alienate it from defence projects initiated by the Conservatives, such as the P1154, HS681 and TSR-2. Therefore such programmes were considered as a waste of public money. However, the government were concerned as the TSR-2 project had already demonstrated to good effect the principles of the demanding operational requirement. As a result the cancellation process had to be shown principally as a condemnation of its costs. At the same time the government had to be seen to be fair to an industry that had, in its time, been responsible for a huge amount of overseas income from aviation exports. The setting up of the Plowden committee, although seen by many as a fair and responsible method of dealing with the problems within the aviation industry, had no power to ensure that its recommendations would be carried out. The committee, allegedly void of expert and unbiased knowledge, appears to have been nothing more than a thinly veiled excuse to justify Labour abandoning the current aviation projects. Denis Healey on a number of occasions admitted that there was a need for an aircraft of TSR-2s capability, even against the background of Britain's withdrawal from the Middle Eastern countries. The Cold War was at a near crisis point and we would require an aircraft that could be stationed in Europe

# CONCLUSION

and would be capable of running front line operations against any Russian incursions into Western Europe. Whilst accepting this situation, Denis Healey still appeared to refuse to recognise that there was perhaps a British solution. However, in his book, *The Time of My Life* he questions why the Buccaneer was never considered, even though, at the time of discussion, George Brown and Roy Jenkins had strongly suggested this particular aircraft.

- It seemed that the industry was taking the blame for past mistakes which, in the majority of cases, had originated within the governments own departments. The effect of this was typified when the government introduced the organisational changes into the industry about which they had little understanding themselves. Although the principles were based on established American methods, there was very little evidence to support the claims that such practices were as efficient as those in use in our own industry, particularly when considering the numbers of aircraft involved. The American industry was based on a much larger scale, producing, in some cases, thousands of one particular aircraft, whilst BAC would have been hard pressed to exceed 100 units. Intervention by the ministry was such that no one appeared fully to understand the consequences of this action. Although the new procedures had been introduced after the lessons learnt from the Supermarine Swift, they were altered to the extent that only the bad points seemed to have survived. This resulted in BAC having to succumb to the changes insisted on by a government in matters about which they, the government, had no previous experience. When the Polaris eventually arrived, the lessons had been learned and that project, using WSM, was a success. As the TSR-2 neared completion, the manufacturers were continuing the programme under the never ending threat of cancellation, hardly the right ingredient to encourage a successful conclusion. When it was seen that the aircraft had achieved most of what had been asked of it, instead of commissioning a full and detailed inquiry involving all the parties concerned to establish that any decision taken was the correct one, the government elected to take the easy way out and just cancelled the project.

- It appeared that a specific course of action had been decided on but was not being adhered to. Both the government and the industry were trying to reach the

*TSR-2 XR220 poses behind some of the design staff.*

same goal, but using different rules and methods and therefore this made the outcome inevitable. This is not to say that we should have carried on regardless, for had the government instigated a specific, in depth review into the affairs of the TSR-2, involving both sides, then a far more positive assessment could have resulted rather than the bungling and interfering which was going on at the time. The industry, ever conscious of cancellation, gave the government a fair option when George Edwards wrote to the minister with 'an enormous incentive', (see Chapter7), whereby the government had very little to lose should they accept. Apparently, ignorant of the success of the programme so far, the government turned the option down as they were not prepared to accept an agreement with open ended responsibility for all future costs in excess of 5 percent. They could hardly have lost as the peak of development had already been achieved.

- We now come to what I believe is the most significant issue in the affair surrounding the TSR-2. The American intervention. There was little doubt that America would have a definite interest in seeing the British aviation industry reduced to such a level of ineffectivness that it would be unable to continue in its present structure. At the same time concern was also being voiced at the threat the British bomber would have on the very markets America was targeting for their own F-111. Therefore America had to devolve a strategy whereby both these issues could be resolved in their favour.

In my opinion, that opportunity came in 1964 when a British delegation visited the USA in order to seek support for the ailing pound. America considered that any form of assistance in the way of credit through the I.M.F, would be the equivalent of supporting the very industry it had most to fear. Therefore America was simply not prepared to sanction supportive measures until Britain gave solid assurances that the Concorde and TSR-2 would be cancelled. Because of a treaty signed previously between Britain and France, the cancellation of Concorde was virtually impossible, however the TSR-2 was a totally different situation.

It was, of course, important that Britian should be seen not have succumbed to American pressure in this matter and so therefore, Britain insisted that the TSR-2 project was being cancelled because its costs were spiralling out of all proportion to its military value. Having cancelled the TSR-2, Britain was then free to order the F-111 under a special agreement.

The F-111 was a product about which even the American Air Force and Navy had considerable reservations. The main sponsor of the aircraft, Robert McNamara, was desperate to secure an overseas customer hoping this would establish confidence in an aircraft that on the face of it appeared to be a failure. Although originally the American plane looked far cheaper, there were conflicting doubts about its performance mainly due to misleading information that had been supplied regarding its specification. In addition there appears to have been no assurances that the cost, specification or delivery of the F-111 would be guaranteed. Indeed, the problems the American aircraft had prior to the cancellation of TSR-2 should have given Britain a clear indication that its costs would inevitably increase. This leads me to the conclusion that, at the time, Britain had little option but to comply with America's conditions and cancel TSR-2, for the simple reason had they not done so America would have witheld its approval of the IMF loan.

# CONCLUSION

*The End. Brand new TSR-2 airframes being scrapped.*
*(Authors Collection)*

Whilst the cost of the project appeared to be high, the demands within the operational requirement would have ensured that Britain had essentially a weapons system capable of carrying out those exacting roles through the 70's and well into the 1990's. Equally, such a weapons system would have been far in advance of anything being considered in the western world and as such would have given both the aircraft and the technology a tremendous export potential.

Not only had Britain's aviation industry suffered well nigh irrepairable damage, but it appeared that the Air Force had taken the brunt of the defence cuts prevalent at the time. Such action seems to indicate that perhaps tactics had been used in an effort to place undue pressure on those in authority, merely to ensure that a strong naval force was retained, irrespective of the logistics.

The Labour governments policy in cancelling, not only TSR-2, but all the major military aircraft programmes had left Britain particularly vulnerable in a period of extreme world political instability. The bombers currently being used by the RAF lacked many of the modern systems associated with TSR-2 and as such would have made them vulnerable to the modern weapons. It seems that Labour had gambled on the countries defences which was akin to accepting a cheap, ineffective insurance policy in the hope that it would not be needed and laying its country open to risk.

Confusion, political backbiting and interference are only part of this sad story. Behind the scenes, within the industry, the urge to get the work done and TSR-2 into the sky was never ending, and neither did those involved at any time feel they could not achieve anything that was asked of them.

# TSR-2 PHOENIX OR FOLLY?

*What could have been adversaries meet at last!*

*TSR-2 and F-111 on the ground at Duxford.*

*(Steve Brooks, IWM)*

During my research for this book I was amazed at the amount of anger and frustration generated by the mentioning of TSR-2. My immediate thoughts were that such emotions would obviously be directed at those who had sought to cancel the TSR-2. However, it became apparent that it was not always as simple as that. In the past Britain's industry had often been accused of 'resting on its laurels' confident in the belief that those wishing to purchase technology would naturally buy British.

The design and building of the TSR-2 represented a huge leap forward in technological advancement and one that would help dispel that myth. Unfortunately, it was the shortsightedness of the Labour government that thwarted the efforts of the industry to prove the sceptics wrong. As it turned out this was to be the last occasion when we could demonstrate that Britain alone was still the world leaders in this field of technology. Although, I believe, we shall continue to show what we are capable of, it will be under the guise of joint efforts where the companies of several nations combine to form a single partnership. Sadly, under this arrangement I fear that the freedom of ideas born out of British initiative risks being strangled out of existence.

Was the TSR-2 a Phoenix or a Folly? We may never know the real truth. So many times the British government has failed to support Britain's industry and we have continually been the laughing stock of all those countries who have gained from our subsequent loss.

Finally, I would like to reflect upon the words of Brian McCann, *"It was a great aeroplane, a threat not only to any potential enemy, but also a threat to the American aviation industry".*

# Appendix I
## Aircrew Flight Log - TSR-2 XR219.

| Date | Flight | Pilot | Navigator | Duration |
|---|---|---|---|---|
| 27th Sep 1964 | 1 | R.P.Beamont | D.J.Bowen | 14 minutes |
| 31st Dec 1964 | 2 | R.P.Beamont | D.J.Bowen | 14 minutes |
| 2nd Jan 1965 | 3 | R.P.Beamont | D.J.Bowen | 8 minutes |
| 8th Jan 1965 | 4 | R.P.Beamont | D.J.Bowen | 20 minutes |
| 14th Jan 1965 | 5 | R.P.Beamont | D.J.Bowen | 22 minutes |
| 15th Jan 1965 | 6 | J.L.Dell | D.J.Bowen | 24 minutes |
| 22nd Jan 1965 | 7 | R.P.Beamont | P.Moneypenny | 28 minutes |
| 23rd Jan 1965 | 8 | J.L.Dell | P.Moneypenny | 27 minutes |
| 27th Jan 1965 | 9 | R.P.Beamont | D.J.Bowen | 22 minutes |
| 6th Feb 1965 | 10 | R.P.Beamont | D.J.Bowen | 29 minutes |
| 8th Feb 1965 | 11 | J.L.Dell | P.Moneypenny | 38 minutes |
| 10th Feb 1965 | 12 | D.M.Knight | P.Moneypenny | 36 minutes |
| 16th Feb 1965 | 13 | J.L.Dell | D.J.Bowen | 45 minutes |
| 22nd Feb 1965 | 14 | R.P.Beamont | P.Moneypenny | 41 minutes |
| 25th Feb 1965 | 15 | J.L.Dell | P.Moneypenny | 72 minutes |
| 26th Feb 1965 | 16 | R.P.Beamont | P.Moneypenny | 47 minutes |
| 8th Mar 1965 | 17 | J.L.Dell | B.McCann | 52 minutes |
| 8th Mar 1965 | 18 | J.L.Dell | P.Moneypenny | 35 minutes |
| 11th Mar 1965 | 19 | J.L.Dell | B.McCann | 33 minutes |
| 12th Mar 1965 | 20 | J.L.Dell | P.Moneypenny | 46 minutes |
| 26th Mar 1965 | 21 | J.L.Dell | B.McCann | 33 minutes |
| 26th Mar 1965 | 22 | J.L.Dell | B.McCann | 35 minutes |
| 27th Mar 1965 | 23 | D.M.Knight | P.Moneypenny | 34 minutes |
| 31st Mar 1965 | 24 | J.L.Dell | B.McCann | 32 minutes |
| 2nd Apr 1965 | 25 | D.M.Knight | P.Monneypenny | Aborted |

These are actual flight times and not from brakes off to brakes on.

## Total Minutes on type

| Pilot | | | Navigator/Observer | | |
|---|---|---|---|---|---|
| R.P.Beamont | = | 245 | D.J.Bowen | = | 198 |
| J.L.Dell | = | 472 | P.Monneypenny | = | 404 |
| D.M.Knight | = | 70 | B.McCann | = | 185 |

# Appendix II
# TSR-2 Production.

| | |
|---|---|
| Prototype Batch | Weybridge |
| Nine Aircraft | Type 571 |
| Ordered: | Oct 1960 |
| Contract No: | KD/2L/02/CB42(a) |
| Build Nos: | K0-1 to K0-9 |
| Serial Nos: | XR219 to XR227 |

| | |
|---|---|
| Prototype Batch | Warton |
| Eleven Aircraft | Type 579 |
| Ordered: | June 28th 1963 |
| Contract No: | KD/2L/13/CB42(a) |
| Build Nos: | K0-10 to K0-20 |
| Serial Nos: | XS660 to XS670 |

| | |
|---|---|
| Pre Production Batch | To be Built at Warton and Weybridge |
| Thirty Aircraft | Type 594 |
| Ordered: | March 20th 1964 |
| Contract No: | KD/2L/16/CB42(a) |
| Build Nos: | Not Known |
| Serial Nos: | XS944 to XS954 |
| | XS977 to XS995 |

# Appendix III
# Contractors associated with the TSR-2 programme.

| | |
|---|---|
| AEI | Generators, Landing Lights, switches, and Electric motors. |
| Aircraft-Marine Prods (G.B) Ltd. | Production of equipment for electrical circuitry and termination for both wiring and instrumentation. |
| Bostik. | Cleaning and Repair compounds. |
| Bristol Siddeley Engines Ltd. | Olympus Power plants. |
| British Aircraft Corporation. | Airframe. |
| British Nylon Spinners. | Working in association with Irving Air Chute to develop special threads able to withstand excesses of heat for braking chute. |
| British Paints Ltd. | Integral Tank Sealants. |
| Cementation (Muffelite) Ltd. | Contributions to Black Boxes. |
| Darlington Chemicals. | Engine Blanket and Heat Sheilds |
| Decca. | Doppler Radar. |
| Delaney Gallay Ltd. | Engine Heat Shields. |
| Dowty Rotol Ltd. | Valves & selectors for Gearboxes. |
| Dunlop. | Pneumatic Jacks, Wheels, Brakes, Maxaret anti-skid units, brake operating and brake cooling systems. |
| DuPont. | Sealing components. |
| Electro Hydraulics Ltd. | Development and production of most hydraulic systems inc main landing gear. |
| Elliott-Automation. | Flight control systems, Head-up-Display, Central Digital Computer. |
| Engins Matra (France) in conjunction with Hawker Siddeley Dynamics (UK). | Martel Missile |
| English Electric. | Automatic scanning ultrasonic testing equipment. |
| English Steel Forge and Engineering. | Structural Components. |
| E.M.I. | Sideways looking radars and Linescan. |
| Ferranti. | Mk4C used in Marwin Max-E-Trace machine.s |
| Ferranti. | Terrain-Following Radar, Inertial Platform and Moving Map Display. |
| F.P.T Industries Ltd. | Rubber Mouldings and Components. |
| P.Frankenstein & Sons (Manchester) Ltd. | Portable Hanger for TSR-2. |
| General Precision Systems Ltd. | Development of Flight simulator. |
| Sir Geo.Godfrey &Partners. | Duplicated air conditioning. |
| Goodyear Tyre & Rubber Co. also Dunlop Tyres. | Tyre development. |
| Graviner. | Fire detection systems. |
| Henry Wiggin & Co Ltd. | Special heat resistant alloys for the turbine inlet. |
| High Duty Alloys Ltd. | Engine and Airframe forgings. |
| H.N.Hobson Ltd. | Power controls. |
| ICI (Silicone Div). | DP47 Hydraulic fluid. |
| IMI Ltd. | Titanium for components in the engines. |
| Imperial Metal Industries (USA) | Aluminium/Lithium alloys for the airframe. |

# TSR-2 PHOENIX OR FOLLY?

| | |
|---|---|
| Irving Air Chute (G.B) Ltd. | Braking parachute systems. |
| Ketay Ltd. | Synchros, servo motors and magnetic amplifiers for various flight control systems. |
| Lucas Gas Turbine Eqpmt. | Engine Intake Control. |
| Lucas. | Hydraulic motor for Rotal alternator. |
| Marconi Co Ltd. | Radio Equipment. |
| Marston Excelsior. | Fuel-cooler heat exchangers. |
| Martin Baker Aircraft Co. | Crew escape systems. |
| Marwin Machine Tools Ltd. | Marwin skin millers and routing machines. |
| M.E.L Equipment Co Ltd. | Electronic units. |
| D.Napier & Sons | Sierraglo lighting for instruments. |
| Normalair Ltd. | Electro pnuematic controls for Air Cond. |
| Normalair Ltd. | Oxygen Breathing installation. |
| Palmer. | Parmatic Filters. |
| Plessey U.K Ltd. | VHF and UHF communication systems. |
| Precision Rubbers. | Specical silicone type seals. |
| Products of GEC (USA). | |
| C.Radyne Ltd. | Heating Eqpmt for Brazing operations. |
| Research Engineering & Controls. | Air Data Sensors. |
| Redifon Ltd. | Data Recorder. |
| Rotax Ltd. | Actuators to control Cameras. |
| Royston Instrumentation Ltd. | MIDAS Data Recorder. |
| Saunders. | Parts of water injection system. |
| Saunders Valve Co Ltd. | Various motorised valves. |
| Sperry Gyroscope Co Ltd. | CL.11 Gyrosyn compass systems. |
| Smiths. | Air data systems, Head-down-Display. |
| S.Smith and Sons Ltd. (Aviation Division). | Navigational, airframe and engine. instruments and Fuel Gauging system. |
| Solar.(USA) | Assisted BSEL with reheat system for the Olympus engines. |
| Standard Telephones and Cables. | F.M Radio Altimeter. |
| Teddington Aircraft Controls Ltd. | Temperature Control Equipment for avionics. |
| Tiltman & Langley Ltd. | High duty terminal switch. |
| Triplex Safety Glass. | Front Screen and quarter lights. |
| Ultra Electronics. | Electrical Throttle system. |
| Vactric Control Equipment Ltd. | Intergrated electronic sub-system components and High Temp servo units. |
| Ward, Brooke and Co Ltd. | Numerous electrical fittings. |
| W.Vinten. | High definition Cameras. |

Source :- Aircraft Engineering November 1964.

# Appendix IV
# Significant Dates in the P17-A / TSR-2 Story.

| | | |
|---|---|---|
| October | 1956 | Preliminary work on a Canberra replacement begins. |
| November | 1956 | Preliminary performance investigation. Podded engines-leading to Rolls-Royce R.B.134. |
| December | 1956 | First low speed wind tunnel tests on a rough model. |
| January | 1957 | 2 x R.B.133 in fuselage. Straight wing. |
| January | 1957 | High Speed Tunnel tests on trapezoidal half-wing. |
| Feburary | 1957 | P.103 Near final configuration. 2 x R.B.133 engines. |
| Feburary | 1957 | E.E.A Report P.103 entitled "Possibilities for a Multi-Purpose Canberra Replacement". Aircraft project No P.17. |
| March | 1957 | Issue of GOR339. (1st Draft) to Ministry of Aviation. |
| May | 1957 | 1/8th Scale full model. (Low speed). |
| June | 1957 | High Speed Tunnel test on delta half-wing. |
| July | 1957 | P.17 larger engines. |
| July | 1957 | OR 3044 (Navigation, Bombing and Reconnaissance and flight Control System). |
| August | 1957 | OR 3596 (Linescan Navigation and Reconnaissance System). |
| August | 1957 | High Speed Tunnel tests on half (1/48th scale model). |
| September | 1957 | Issue of GOR 339 to Industry. Submissions invited from Blackburn and General Aircraft Ltd. Bristol Aircraft. De Havilland. Fairey Aviation. Handley Page. Short Bros. Vickers-Armstrong Hawker Siddeley English Electric. |
| October | 1957 | First examination of P.17-D. |
| October | 1957 | P.17 Vertical Take-off study. |
| Oct/Nov | 1957 | English Electric - Shorts collaboration begins. |
| November | 1957 | Joint English Electric Co - Shorts programme. |
| November | 1957 | Half model - blown flaps. |
| December | 1957 | Shorts test P.17-D. |
| January | 1958 | Brochure produced. |
| February | 1958 | O.R.339 - 1st Draft. |
| March | 1958 | Report made on the selection of companies. |
| April | 1958 | O.R.339 - 1st Issue. |
| April | 1958 | Exchange of views with V-A on equipment. |
| June | 1958 | Final Ministry of Aviation assesment. |
| July | 1958 | High Speed tunnel tests on complete model (1/64th Scale). |
| July | 1958 | Quarter-Scale intake model sent to Rolls-Royce. |
| August | 1958 | New low speed model. |
| August | 1958 | O.R.339 - 2nd Draft. |
| September | 1958 | Quarter-Scale intake model sent to Bristol's. |
| November | 1958 | O.R.339 - 3rd Draft. |

## TSR-2 PHOENIX OR FOLLY?

| | | |
|---|---|---|
| November | 1958 | First joint discussion at Technical level with Vickers-Armstrong at Weybridge. |
| December | 1958 | O.R.339 - 4th Draft. |
| December | 1958 | Bristol Siddeley Engines chosen. |
| January | 1959 | Parliamentary statement by Minister of Supply that V-A and E. E. A. are to undertake the development of the new plane to be known as the TSR-2. |
| February | 1959 | Joint V-A / E.E.A design team formed at Weybridge. |
| March | 1959 | O.R.343 - Final Draft. |
| April | 1959 | Design study contract V-A / E.E.A. Discussions begin. |
| May | 1959 | O.R.343 - 1st Issue. |
| June | 1959 | Design study contract 01 (Covering January.59 to Dec.59). |
| June | 1959 | V-A Purchase order on E.E.A (WX 6701). |
| July | 1959 | V-A / E.E.A Preliminary Brochure. |
| October | 1959 | E.E.A Design Team return to Warton from Weybridge. |
| March | 1960 | Draft Contract to Vickers-Armstrong from the Ministry. |
| June | 1960 | Interim Design Study Contract 04. (Covering January.60 to Sep.60). |
| June | 1960 | V-A Purchase order on E.E.A (WF 4043A). |
| August | 1960 | Draft Contract for 9 development batch aircraft received. |
| August | 1960 | Specification RB.192D |
| October | 1960 | Contract for 9 development batch aircraft issued. (KO-1 to KO-9). Contract No KD/2L/02/CB42(a). |
| November | 1960 | E.E.A internal contract issued (9 development batch aircraft). |
| November | 1960 | V-A Purchase order on E.E.A. |
| February | 1962 | Firing of the 1st Free Flight model at Aberporth. |
| February | 1962 | Spinning model tests at Larkhill. |
| June | 1963 | Contract for 2nd Batch (KO-10 to KO-20). Contract No KD/2L/13/CB42(a). |
| March | 1964 | Contract for Material for Aircraft 21 - 50. |
| September | 1964 | First Flight from Boscombe Down. |
| February | 1965 | 1st Prototype arrives at Warton. |
| April | 1965 | Announcement by J.Callaghan of the Cancellation of the TSR-2 in the Budget Speech. |
| June | 1965 | Turning down of proposal to use TSR-2 for research work. |
| July | 1965 | Closure of TSR-2 contract. |

Source :- British Aerospace.

# Appendix V

## Preserving what's left...

Significant parts of the TSR-2 are preserved around the country. Two airframes are with major museums, and other items are around with other collections. All are members of the overall co-ordinating organisation...

# British Aviation Preservation Council

There is a certain irony in the fact that TSR-2 is now a 'museum piece', for when it was born around 1960 Britain did not have a significant aviation museum. The Science Museum and the Imperial War Museum both displayed some aircraft as part of their comprehensive themes and there was the Shuttleworth Collection at Old Warden - but no Royal Air Force Museum, no Fleet Air Arm Museum and no Museum of Army Flying. The total number of preserved aircraft was probably around 100.

During the short life-span of the TSR-2 the whole picture changed, for there was a sudden surge of interest in aircraft preservation - led not by the museum professionals, but by enthusiastic amateurs. One of the first of the new museums was the Mosquito Aircraft Museum at Salisbury Hall, Hertfordshire, which opening in 1959 to display the prototype Mosquito at the location where it was designed. Other museums followed and, in 1967, the British Aircraft Preservation Council was formed by ten organisations, of which eight were ran by volunteers. The other two were the newly formed Royal Air Force Museum and the Fleet Air Arm Museum.

When BAPC celebrated its tenth anniversary, it had 48 full member organisations preserving about 500 aircraft, plus 10 associate members. Membership of BAPC has always been restricted to organisations directly, or indirectly involved in preservation. Individuals are encouraged to join one of the member organisations in their locality. As the number of preserved aircraft grew, so did the interest in other aspects of aviation preservation; airfields, radio, radar, weapons, uniforms and vehicles. This in turn led to a change of name (but not initials) for BAPC, which became the British Aviation Preservation Council. By its 25th Anniversary in 1992, membership had grown to over 100 full members and 30 associate members, including overseas members in Europe, America and Australasia. The current total of preserved aircraft in Great Britain stands at around 1,000.

This figure is unlikely to grow dramatically in the foreseeable future as most of the old relics stored in barns have been retrieved, and the number of aircraft being taken out of service has fallen away. The problem now is to conserve the surviving aircraft, particularly as about half of them are kept in the open air. A significant number of preservation societies from the early days of BAPC have evolved into fully fledged museums. Their recent efforts have been directed towards building hangars for their aircraft and facilities for their visitors. All TSR-2 items are held in BAPC member museums - be them complete airframes at Cosford and Duxford or other items at Weybridge or Coventry - in environments that ensure long-term preservation.

As a co-ordinating body, BAPC continues to act as a forum, and the quarterly meetings provide an opportunity for members not only to discuss problems, but also to make known their wants and disposals. BAPC is also tackling the problems of conservation with two initiatives. Preserved aircraft are being 'listed' (just as important buildings are listed) so that limited resources can be directed to the most deserving cases. The second project is to organise a series of conferences under the banner title 'Stopping the Rot'. to help member organisations with the nitty-gritty problems of conservation and restoration.

Don Storer
Secretary, BAPC

# THE AEROSPACE MUSEUM

Although the concept of a National Aviation Museum was put forward as long ago as 1917, it was not until 1972 that such a proposal came to fruition following Dr John Tanner's paper presented to a Board, chaired by Marshal of the Royal Air Force Sir Dermont Boyle recommending that such a museum be established. At the time neither the money or suitable site existed, but after intensive negotiations with the Treasury, it was agreed that whilst funds would be made available to run such an establishment, capital expenditure would have to be met by private funds.

The museum came into being at RAF Hendon during the late 1960's and was officially opened to the public in 1972. Within five years the museum had expanded to include the Battle of Britain and Bomber Command Museums. It was realised that Hendon would not be able to house the extensive collection of aircraft and memorabilia on one site, therefore an additional collection was formed at RAF Cosford near Wolverhampton in 1979. Originally the collection was only open at weekends, but, because of increasing popularity it was decided to keep it open all year round except on 1 January and 24-26 December.

The museum is situated within the RAF Station itself, and although the hangars, grounds and most of the exhibits belong to the Ministry of Defence, the Museum itself receives no public funds and is a registered charity relying solely on gate receipts and public donations.

The day-to-day running is looked after by just eight full-time staff, but receives assistance in maintaining the exhibits from the friends of the Museum, the Aerospace Museum Society, which has in excess of 300 members, although only around 40 are actively involved in the cleaning, painting and restoring of exhibits under the supervision of the museum's Technical Curator, Bill Roseby.

In all there are over 70 aircraft, representing a period from World War II to the present day. Alongside the unique collection of captured German warplanes are some unusual British aircraft which were used during the war. The latest edition is a Buccaneer S2 which saw action in the Gulf conflict and retains the desert colours and nose artwork.

Alongside the aircraft exhibits is a rare collection of rockets and missiles, many of which are representative of the types used by Germany during the Second World War. The Museum also has a large display of aero engines which range from the small piston engine type to the huge Rolls Royce RB211 jet engines as used in the Lockheed L-1011 Tristar and Boeing 747. The comprehensive transport display was made even more so by the addition of a large collection of equipment donated by British Airways, which traces the history of Britain's national airline. The airline also donated a number of airliners in the various liveries of British Airways or its predecessors.

Cosford is also unique in that it has a hangar dedicated to many examples of post-war research and development aircraft. These include such forgotten names as the Gloster Meteor, Fairey Delta II, Bristol 188 and Saunders-Roe SR53. Forming part of this collection is the TSR-2, which was delivered to the museum in 1975 from the site at RAF Henlow known as 'The Pickle Factory'.

The airframe stood for three years before restoration work began in 1978. Spearheaded by Tom Thomas, Chairman of the Aerospace Museum Society, the work initially consisted of cleaning and tidying up XR220 before work began in earnest on a major restoration programme in 1979. In 1990 the Museum received an unexpected bonus when a number of containers arrived marked 'TSR-2' After sorting through the contents, enough equipment was found to warrant an attempt to re-build the front and rear cockpits as well as the avionics bay as it would have looked on the development aircraft. They also fitted both cockpits with the original Martin-Baker ejection seats.

The Museum was presented with a number of Bristol Siddeley Olympus 320 engines as used in the TSR-2 and one has been sectioned in order to illustrate the workings of a twin-spool axial engine.

In 1991 a reunion of TSR-2 Flight Test crew was organised and Jimmy Dell, (BAC Chief Test Pilot) Don Knight (Deputy Chief Test Pilot) and Brian McCann (Navigator/Observer) visited the museum to reflect on the state of aviation since TSR-2. Jimmy Dell also presented the museum with the control stick from XR219.

Today the museum is justly proud of not only its TSR-2, but also of the many other rare and unique exhibits as well as the comprehensive facilities it is able to offer its thousands of visitors annually.

# Imperial War Museum DUXFORD

In the early 1970's the Imperial War Museum of London were seeking a suitable location where they could store and display some of their larger exhibits that could not be easily accommodated in the main building in their Lambeth Headquarters. The site had to be reasonably close to London with easy access by road. At the same time the Home Office were considering what to do with the Royal Air Force station of Duxford in Cambridgeshire which was surplus to requirements. Thus a new and facinating living museum facility came about.

The airfield, which has a long and famous history, had ceased to be operational in 1961, but was not declared surplus to requirements until 1969. Built in 1917, the airfield became well known during the Second World War, when it played a vital role in the Battle of Britain. Later in the war Duxford became host to the 78th Fighter Group 8th Air Force USAAF with their P-47 Thunderbolts and later P-51 Mustangs.

After the war the airfield was again occupied by the Royal Air Force and the familiar sound of the Spitfire was heard again over Duxford. But not for long, for the purr of Merlins was replaced with the unfamiliar whine of the jet engine in 1946 when the Gloster Meteor arrived. By 1949 the PSP/grass runway was declared unsuitable for jet aircraft, so the airfield was placed on Care-and-Maintenance whilst the new concrete runway was laid. Although the first aircraft to return to Duxford was the Meteor F8, the station also hosted various squadrons with a variety of aircraft, including the Hawker Hunter and Gloster Javelin before it closed in 1961.

In 1967 Spitfire Productions received permission to use the airfield for making the film *'The Battle of Britain'*, spending some £38,000 on tidying up the base and its buildings before filming began. The Ministry of Defence eventually decided that they no longer had any use for Duxford and made the site available to interested parties. Numerous proposals were put forward and turned down after a public enquiry in 1971. Meanwhile the Imperial War Museum had begun to use one of the hangars to restore a P-51D *Mustang* with the help of the East Anglian Aviation Society. As the exhibits grew in numbers, Duxford put on an air display in October 1973; by 1975 the Imperial War Museum had received permission to make use of the whole site and, in 1976 began opening the public daily.

The Museum has a number of unusual and unique exhibits that include the only examples of a B-29 *'Superfortress'* and a B-52 *'Stratofortress'* in the UK as well as the only F-111 swing-wing bomber in a British museum. Duxford also has a fine collection of civil airliners, including the British Pre-Production Concorde 01 which flew in to Duxford in 1977, just before the construction of the M11 motorway curtailed the usable runway length.

Duxford's TSR-2, XR222, was presented to the Museum by the Ministry of Defence in 1977 and was delivered to the site in 1978. The airframe, which was the fourth on the Weybridge Production line and was incomplete at the time of cancellation, had been previously used at the Cranfield Institute of Technology for post graduate teaching. The Institute had been fortunate to receive such an aircraft but, because the aircraft was to spend many years in the open air with many of the access panels missing, corrosion was beginning to have an adverse effect.

In 1984 a team of Instructors and Apprentices from BAe Preston manufactured a number of the missing panels, but a major restoration programme was not formulated until 1988. This mammoth task took members of the Duxford Aviation Society, led by IWM staff, just over a year to bring the aircraft up to display standard, and included a complete re-spray of the airframe. The Imperial War Museum decided that because of the lack of equipment, (that had never been fitted originally) they would be unable to restore the cockpits to a display standard. In February 1990, Roland Beamont was invited to the Museum to inspect the aircraft before it was moved to the display area.

Today the site at Duxford has perhaps one the the largest collections of both flying and static exhibits of significant civil and military aircraft in Europe and regulary holds functions and air displays. The site is open all year round except 24-26 December.

For more information call 0223 835000 of the Events Hotline on 0891 516816.

and finally...
Seen in February 1990 following its restoration to display standard, XR222 is parked behind some of the workers. Back Row, (L to R) D Fisher, P Prest, A Robinson, M Oxley, R Greenwood, C Stead, J Hoath, A Hoath, M Howard, T Minshal, D Lander. Front Row, (L to R) Bevis Griffiths, David Lee, Roland Beamont, Chris Chippington, wearing a custom-built TSR-2 jumper is the author Frank Barnett-Jones and Eric Perrott